Molecular Protocols in Transfusion Medicine

T0229046

Molecular Protocols in Transfusion Medicine

Gregory A. Denomme, PhD

Assistant Scientist, Pathology and Laboratory Medicine,
Mount Sinai Hospital, and The Canadian Blood Services,
Toronto

Maria Rios, PhD

Manager, Science and Technology Laboratory
New York Blood Center, New York

Marion E. Reid, PhD

Head, Immunochemistry Laboratory
New York Blood Center, New York

ACADEMIC PRESS

A Harcourt Science and Technology Company

San Diego   San Francisco   New York   Boston
London   Sydney   Tokyo

Copyright © 2000 by ACADEMIC PRESS

Academic Press
A Harcourt Science and Technology Company
32 Jamestown Road, London NW1 7BY, UK
http://www.academic press.com

Academic Press
A Harcourt Science and Technology Company
525 B Street, Suite 1900, San Diego, California 92101-4495, USA
http://www.academic press.com

ISBN 0-12-209370-4

A catalogue for this book is available from the British Library

Typeset by Phoenix Photosetting, Chatham, Kent
Printed and bound in the United Kingdom
Transfered to Digital Printing, 2011

CONTENTS

PREFACE

The understanding of the genetic basis for blood group systems has expanded enormously over the past 5 years. This manual is a compendium of genetic polymorphisms for blood group antigens relevant for transfusion medicine and the techniques to detect them. We encourage comments from readers on errors, omissions, and suggestions for consideration for inclusion in the next edition. Please write to the Editor, Molecular Protocols in Transfusion Medicine, Harcourt Publishing Limited, Harcourt Place, 32 Jamestown Road, London NW1 7BY, UK, or to Marion Reid, Immunochemistry Laboratory, New York Blood Center, 310 East 67th Street, New York, NY 10021, USA (E-mail: mreid@nybc.org).

We are indebted to Christine Lomas-Francis and Jill Storry for critically reading the manuscript and making valuable suggestions. We appreciate the enthusiasm and help from Tessa Picknett and Lilian Leung. We are indebted to Robert Ratner for preparing the manuscript and figures.

LIST OF ABBREVIATIONS

ACD	Acid citrate dextrose
A, C, G, T	Adenine, cytidine, guanidine, thymidine nucleotides
AS-PCR	Allele-specific polymerase chain reaction
CAP	College of American Pathologists
cGLP	Current good laboratory practice
DAT	Direct antiglobulin test
dd H_2O	Deionized distilled H_2O
dNTP	Deoxyribonucleotide triphosphates
DNA	Deoxyribonucleic acid
EDTA	Ethylenediamine tetraacetic acid
FBS	Fetal bovine serum
FDA	Food and Drug Administration
HDN	Haemolytic disease of the newborn
HLA	Human leukocyte antigen
HPA	Human platelet antigen
ISBT	International Society of Blood Transfusion
MSDS	Material Safety Data Sheets
NA	Neutrophil antigen
NATP	Neonatal alloimmune thrombocytopenic purpura
OD	Optical density
PBS	Phosphate buffered saline
PCR	Polymerase chain reaction
RBC	Red blood cell
RFLP	Restriction fragment length polymorphism
RNA	Ribonucleic acid
RPMI	Roswell Park Medical Institute (Tissue Culture Media)
SD	Standard deviation
STR	Short tandem repeat
TAE	Tris-acetate EDTA
TEMED	N,N,N',N'–Tetramethylethilenediamine

TBE	Tris-borate EDTA
TRALI	Transfusion-related acute injury
U	Unit
UV	Ultraviolet
URL	Uniform Resource Locator
VNTR	Variable number of tandem repeat
WBC	White blood cell

Part 1   Introduction

1.1   Background

Since the development of the polymerase chain reaction (PCR) (Mullis and Faloona 1987), a plethora of applications using this technique have been reported. The majority of genes encoding either proteins that carry blood group antigens or transferases that attach blood group-specified carbohydrates have been cloned and the molecular bases associated with many cell surface markers on red blood cells (RBCs), platelets, and neutrophils have been determined. Thus, it is possible to perform molecular analysis for these inherited alleles. Such analyses include allele-specific PCR amplification (AS-PCR), and PCR amplification followed by restriction enzyme digestion (PCR-restriction fragment length polymorphism (RFLP)).

There are 25 red cell blood group systems, consisting of over 200 antigens, which are recognized by the International Society of Blood Transfusion (ISBT) (Daniels et al 1995, 1996, 1999). Genes encoding 21 of these systems have been cloned and sequenced. The molecular basis that gives rise to many of the polymorphisms that result in blood group antigens and phenotypes are known (Reid and Lomas-Francis 1996; Reid and Yazdanbakhsh 1998; Avent 1997).

There are 14 human platelet antigen systems that are expressed on seven glycoproteins. Six of the seven genes have been cloned and sequenced (Warkentin and Smith 1997; Santoso and Kiefel 1998).

There are eight neutrophil systems consisting of at least 11 antigens. Three of six genes have been cloned and sequenced. These antigens are not related to human leukocyte antigens (HLA) expressed on neutrophils.

A wide variety of molecular events generate blood group antigen diversity (Reid and Rios 1999; Alberts et al 1994). These include:

- missense mutation (by far, the most common)
- nonsense mutation
- mutation of motifs involved in transcription
- alternative splicing
- deletion of a gene, exon(s) or nucleotide(s)
- insertion of an exon(s) or nucleotide(s)
- alternate initiation
- chromosome translocation
- single crossover
- gene conversion
- other gene rearrangements

To provide a comprehensive clinical service, the following skills are desirable when interpreting results of molecular analyses:

- a basic knowledge of the principles and mechanisms of molecular techniques
- an understanding of gene structure, function, and expression
- an understanding of blood group antigens and the serological techniques used to characterize their expression
- an ability to correlate results to the clinical problem being addressed

1.2   Aim, Scope and Purpose of the Book

The aim of this book is to provide general principles, generic protocols for basic techniques, and specific protocols for genotyping red cell, platelet and neutrophil blood groups in clinical or research transfusion medicine laboratories. The protocols we provide for specific alleles have been shown by us to be robust and reproducible in a clinical setting. For other alleles, protocols are not given but relevant information is provided on the relevant gene figure or in a table. The book is designed in a convenient, easy-to-use format with pertinent references and protocols for performing molecular analyses.

We have made the assumption that the reader has a basic understanding of the genetic code, and of the pathway from a gene to its product (Figure 1.1). Such information is found in molecular biology textbooks (Alberts et al 1994) and review articles on blood group genetics (Reid and Rios 1999; Avent 1997; Daniels 1999; Warkentin and Smith 1997;

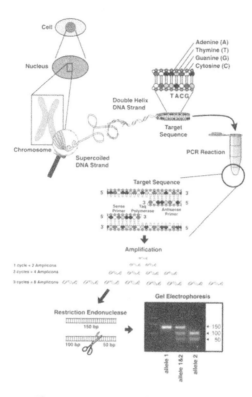

Figure 1.1   From cell to PCR RFLP.

Stroncek 1997). Details for the design of primers, and gene sequence analysis, are also beyond the scope of this book. Description of these and other more complex techniques can be found elsewhere (Sambrook et al 1989; Ausubel et al 1997).

1.3   Organization of the Book

The book is divided into seven parts. Part 1 provides a brief overview of the background, aims and scope of the contents; Part 2 gives general considerations and principles for molecular analyses; Part 3 gives molecular protocols including preparation and storage of DNA, preparation of gels and reagents, PCR-RFLP, AS-PCR and quality control;

Parts 4–6 give terminology, clinical applications, facts sheets, and molecular protocols; Part 7 contains Uniform Resource Locator (URL) addresses and a brief description of databases that are available through the Internet, a list of definitions, and examples of worksheets and forms.

Facts sheets give information about the gene name, chromosomal location, size of DNA and RNA, product name, exon and intron arrangement with location of alleles (Figure 1.2), selected GenBank accession numbers and prevalence of common polymorphisms. The molecular protocols give the nucleotide substitutions for selected clinically relevant polymorphisms (as indicated on the gene map), location of primers, size of PCR amplicon, size of expected fragments after restriction enzyme digestion, and primer sequences. The PCR protocols and amplification profiles are referred to by the numbering system given in Part 3. Information about alleles and relevant restriction enzymes for which assays have not been developed in our laboratories are provided in tabular form. Key references are provided.

Information in the databases, together with the principles outlined within this book, should allow investigators to establish their own protocols.

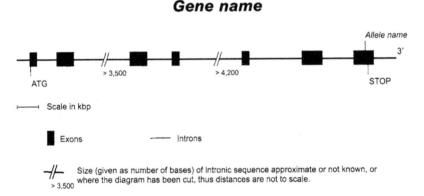

Figure 1.2   Proportional diagram of a gene showing organization and size of exons and introns. Exons counted sequentially from 5′ to 3′.

1.4   Terminology

Throughout this book we refer to AS-PCR and PCR-RFLP as 'genotyping'. However, it must be remembered that only a single nucleotide of a particular gene is being defined. In order to determine the true genotype it would be necessary to sequence the whole gene. Throughout this book we use the term 'genotype' to denote the polymorphism determined by molecular analysis on DNA. We use 'phenotype' to denote the expressed antigen detected with a specific antibody.

The terminology used for genes, alleles and antigens is that recommended by the ISBT (Daniels et al 1995, 1996, 1999; von dem Borne and Decary 1990). In some instances, the Human Gene Mapping (HGM) name for a gene (Antonarakis et al 1998) may also be used; such use will be indicated in parenthesis. A gene is written in italics, e.g., *FY*; *GP3A*. An allele is written in italics with a space followed by a number or letter, e.g., *FY 1*; *GP3A 1*. If italics are not used, an asterisk is placed between the gene symbol and allele symbol, e.g., FY*1; GP3A*1. Nucleotide changes are given after the number in cDNA, e.g., 125G>A. Amino acid changes are given either side of the residue number, e.g., G42D. Numbering starts with 1 at the initiation codon or 1 of the mature protein, such as when the encoded product has a leader sequence that is cleaved after insertion into the membrane.

1.5   Regulatory Compliance

No specific FDA regulations for molecular analysis currently exist. However, current good laboratory practice (cGLP) should be applied and the regulations of the Occupational Safety Health Authority and the Chemical Hygiene Plan should be followed. All specimens should be handled in accordance with the Blood Borne Pathogen recommendations using universal handling precautions. Material Safety Data Sheets (MSDS) should be readily available for all reagents. In addition, the development of a molecular diagnostic laboratory requires a number of policy, procedural and operational considerations. In this regard, the College of American Pathologists (CAP) has established a Laboratory Accreditation Program. Many elements of the CAP molecular pathology inspection checklist (Laboratory Accreditation Program, College of American Pathologists, Northfield, IL) are applicable to the molecular analysis of blood group alleles. Also the CAP recommends that procedure manuals be written in compliance with the National Committee for Clinical Laboratory Standards (*Clinical Laboratory Technical Procedure Manuals*, 3rd Edition, 1996, Vol. 16, No. 15,

National Committee for Clinical Laboratory Standards. Wayne, PA). The procedures and protocols written in this manual comply with those recommendations.

When performing PCR analyses one should be aware of the limited licence agreement for the use of *Taq* polymerases and the PCR technique. The Roche Molecular Systems Inc. (AmpliTaq Gold™, Product Monograph, Roche Molecular Systems, Inc., Alameda, California, 1996) licence states 'No right to perform or offer commercial services of any kind using PCR, including without limitation reporting the results of purchaser's activities for a fee or other commercial considerations, is hereby granted by implication or estoppel'.

In the absence of regulatory standards and proficiency samples, it is important to have validation data and provision for semi-annual evaluation of each procedure. Each laboratory performing molecular analyses should write a Quality System that includes the following essentials based on ISO 9000 recommendations:

- management responsibility/organization
- personnel
- equipment
- supplier issues
- agreement/contract review
- standard operating procedures
- process control
- documents, records and data control
- incidents, errors and accidents
- assessments
- process improvement
- facilities and safety

1.6   References

Alberts, B., Bray, D., Lewis, J., et al (1994) *Molecular Biology of the Cell*, 3rd edition. Garland Publishing, New York.

Antonarakis, S.E. and Nomenclature Working Group (1998) Recommendations for a nomenclature system for human gene mutations. *Human Mutation* 11, 1–3.

Ausubel, F.M., Brent, R., Kingston, R.E., et al (1997) *Current Protocols in Molecular Biology* (3 volumes). John Wiley & Sons, New York.

Avent, N.D. (1997) Human erythrocyte antigen expression: its molecular bases. *Br J Biomed Sci* 54, 16–37.

Daniels, G. (1999) Functional aspects of red cell antigens. *Blood Rev* 13, 14–35.

Daniels, G.L., Anstee, D.J., Cartron, J.-P., et al (1995) Blood group terminology 1995. ISBT working party on terminology for red cell surface antigens. *Vox Sang* 69, 265–279.

Daniels, G.L., Anstee, D.J., Cartron, J.P., et al (1996) Terminology for red cell surface antigens – Makuhari report. *Vox Sang* 71, 246–248.

Daniels, G.L., Anstee, D.J., Cartron, J.P., et al (1999) Terminology for red cell surface antigens – Oslo report. *Vox Sang,* 77, 52–57.

Mullis, K.B. and Faloona, F.A. (1987) Specific synthesis of DNA in vitro via a polymerase-catalyzed chain reaction. *Methods Enzymol* 155, 335–350.

Reid, M.E. and Lomas-Francis, C. (1996) The Blood Group Antigen FactsBook. Academic Press, San Diego.

Reid, M.E. and Rios, M. (1999) Applications of molecular genotyping to immuno-haematology. *Br J Biomed Sci,* 56, 145–152.

Reid, M.E. and Yazdanbakhsh, K. (1998) Molecular insights into blood groups and implications for blood transfusions. *Current Opinion in Hematology* 5, 93–102.

Sambrook, J., Fritsch, E.F. and Maniatis, T. (1989) *Molecular Cloning: A Laboratory Manual* (3 volumes), 2nd edition. Cold Spring Harbor Laboratory Press, Cold Spring Harbor, NY.

Santoso, S. and Kiefel, V. (1998) Human platelet-specific alloantigens: Update. *Vox Sang* 74 (Suppl 2), 249–253.

Stroncek, D.F. (1997) Neutrophil antibodies. *Current Opinion in Hematology* 4, 455–458.

von dem Borne, A.E. and Decary, F. (1990) ICSH/ISBT Working Party on platelet serology: Nomenclature of platelet-specific antigens. *Vox Sang* 58, 176.

Warkentin, T.E. and Smith, J.W. (1997) The alloimmune thrombocytopenic syndromes. *Transf Med Rev* 11, 296–307.

Part 2   General Considerations

2.1   Clinical Information

The genotype may not reflect the phenotype when the patient has undergone one of certain medical procedures. Transplantation with homologous stem cells and recent massive or chronic transfusion can result in apparent disagreement. In the prenatal setting, a history of artificial insemination, *in vitro* fertilization, questionable paternity, or surrogate motherhood can complicate the interpretations. It is helpful to obtain the ethnic background of the patient. Gene frequencies and the presence of rare variant alleles differ among ethnic groups, which impacts on the analysis.

2.2   Basic Principles of Molecular Analysis

2.2.1   Source of DNA

White blood cells (WBCs) are the most common and convenient source for DNA. However, because genomic information is analysed, DNA can be obtained from any nucleated cell. Non-invasive sources include epithelial cells obtained from the urinary tract (urine sample) or mouth (buccal swab) (Rios et al 1999). Such sources are reasonable alternatives when a blood sample is difficult to obtain.

2.2.2   Quality, Quantity and Purity of DNA

The quality, quantity or purity of the extracted DNA can affect the efficiency of PCR amplification. Therefore, it is advisable to assess these parameters prior to analysis. Protocols to assess DNA quality, quantity and purity are provided in Part 3. Samples provided for molecular analysis may not be ideal, e.g., frozen blood droplets or cells from amniotic fluid.

Degraded DNA may consist of fragments that are smaller than the target sequence. The effect of degraded DNA can be overcome by the amplification of shorter PCR products (<150 bp). DNA degradation is assessed by electrophoresis in an agarose gel followed by staining with ethidium bromide: a smear-like appearance indicates poor quality. DNA contaminated with protein (i.e., carryover during the extraction process) is generally not suitable for PCR. The purity of the extracted DNA can be assessed by optical density measurement at 260 nm and 280 nm. Pure DNA has a 260/280 ratio of 1.9 ± 0.1. Protein contamination is suspected when a lower than expected ratio is obtained (Sambrook et al 1989). When extraction problems do occur, the most common contaminant from whole blood is haemoglobin.

It is important to determine the DNA concentration since the amount of starting template can affect the yield and specificity of the PCR. The amount of DNA required for each PCR assay is dictated by the size of the fragment to be amplified. Strategies for the amplification of large fragments require more template, especially for DNA of poor quality. However, excessive amounts of template will result in absent, poor, or non-specific amplification. As a general rule, 50–200 ng of DNA are sufficient for fragments of ~150 bp, whereas 200–500 ng may be required for targets from 250 to >1,000 bp.

2.2.3 Primer Design

Primers are short sequences of single-stranded oligonucleotides designed to anneal with a specific region of DNA. Ideally, primers should consist of 18 to 22 nucleotides and have a G+C content between 50% and 60%. Primer pairs should not contain internal homology nor have sequences that form secondary structures. A primer should not have three or more consecutive G or C nucleotides at the 3′ end (Dieffenbach et al 1993). Taking these considerations into account should result in primers with annealing temperatures between 55°C–62°C. A convenient formula to estimate the annealing temperature of a primer is:

$$T_m°C = 2(A+T) + 4(C+G)$$

When reconstituted at 20×10^{-6} M, each microlitre will contain 100–200 ng, which is sufficient to amplify 50–500 ng of target DNA. The following equation is used to calculate concentration where 330 is the average molecular weight of a deoxynucleotide, N is the primer length, C is the desired concentration (mol/L), and 1,000 is the conversion factor (g/L to ng/µL):

$$ng/µL = 330 \times N \times C \times 1,000$$

If possible, primers should be located in exonic sequences, since intronic regions are prone to more variation, which may affect the PCR. If an intron region primer is unavoidable, it must be tested as extensively as possible using DNA from various ethnic populations.

Table 2.1

Molarity of primer based on length and concentration (ng/µL)

Primer length	Concentration (pmol/µg)	Molarity (20 µM)
18-mer	168	119 ng/µL
20-mer	152	132 ng/µL
25-mer	121	165 ng/µL
30-mer	101	198 ng/µL

2.2.4   DNA Polymerases

Taq DNA polymerase is the most common enzyme used for PCR amplification. This enzyme is extremely heat resistant with a half-life of 40 minutes at 95°C. At its optimal temperature (72°C), nucleotides are incorporated at a rate of 2–4 kilobases per minute. However, polymerases are active over a broad temperature range (Gelfand and White 1990). Therefore, if DNA, primers and *Taq* polymerase are mixed together, mis-priming and elongation can occur at room temperature (before the denaturation step). This will result in the amplification of non-specific targets that can be overcome by the use of a 'hot-start' PCR technique (Mullis 1991). Typically the technique involves either: (1) adding the polymerase during the initial denaturation step; (2) separating the enzyme from the other reagents using a paraffin layer (AmpliWax™; Roche Molecular Systems, Inc.); or (3) using a modified enzyme that is activated upon heating to 95°C for 15 minutes. In each instance, polymerase activity is prevented until the initial annealing stage.

Taq polymerase does not have 3'–5' proof-reading activity. However, it has a very low mis-incorporation rate; estimated at 1 per 10,000 bases. Proof-reading enzymes must be avoided when performing AS-PCR, since these enzymes will correct the deliberate mis-match necessary for genotyping.

2.2.5   Amplification Efficiency

In addition to the quality, purity and quantity of DNA, the efficiency of PCR amplification is directly affected by a combination of factors including the concentration of reagents ($MgCl_2$, primers, enzyme and enzyme buffer), and the PCR amplification profile (number and conditions of cycles) (Cha and Thilly 1993). In order to control PCR amplification efficiency and to verify whether the PCR reagents are working properly, each reaction should include an internal control. AS-PCR requires the amplification of an unrelated target using an independent pair of primers since allele-specific amplification may result in no PCR product. The amplification of an irrelevant target validates that all reagents and target DNA were added to the reaction mixture and indicates absence of inhibitory substance(s). In contrast, PCR-RFLP uses a single set of primers that flank the polymorphic region and amplifies both alleles, therefore, dispensing with the need for unrelated internal control for amplification. However, ideally, the amplified fragment should include a control restriction enzyme site in all targets to monitor enzyme performance. It is critical to optimize PCR conditions and amplification profile for each analysis. Such optimization is required even when small changes are made, e.g., changes in the reagent supplier, and implementing a published protocol.

2.2.6   Amplification Accuracy

Although cross-contamination of blood samples by previous analysis on clinical automated machines and microchimerism in blood samples from massively transfused trauma patients have been observed (Reed et al 1997; Bianchi et al 1996), the PCR studies in these reports were designed intentionally to detect rare sequences. For genotyping transfused patients, the quantity of transfused WBCs in the specimen is insufficient to interfere with the recipient's genotype (Reid et al 1999; Wenk and Chiafari 1997).

2.2.7   Avoiding Contamination

The majority of contamination is due to pre-amplified products that are carried into DNA samples or into any of the reagents. Carryover contamination is far more serious than transfusion-induced microchimerism, since contaminating DNA templates are preferentially amplified over genomic DNA in the PCR assay. The PCR is prone to contamination either from products of a previous reaction or from aerosols that contaminate subsequent reactions or reagents. Contamination of PCR-based assays can be prevented by taking the following simple precautionary measures (Cimino et al 1991; Dragon 1993; Fahy et al 1994; Kwok and Higuchi 1989; Prince and Andrus 1992):

- Areas for sample handling, pre-amplification PCR set-up, and post-amplification analysis should be physically separated.
- Equipment, reagents, solutions and laboratory coats must not be shared among the areas.
- Gloves should be changed frequently to reduce the possibility of contamination.
- All tubes must be nuclease free. Screw caps are preferable for PCR reagents and primers since snap caps tend to generate aerosols upon opening.
- Barrier tips should be used to reduce micropipettor contamination and aerosol carryover.
- Reagents used for PCR must be frozen in aliquots. The 'in use' aliquots must be discarded if contamination is suspected.
- All equipment used for PCR analyses must periodically be wiped with a 1:10 dilution of bleach, e.g., racks, micropipettors, centrifuges. This is superior to the use of UV light decontamination that works by cross-linking DNA. As contamination is caused by short DNA sequences (amplicons of 1 kb or less) the chance of effectively cross-linking every molecule of such length is remote.

2.2.8   Interfering Substances

Some substances, e.g., heparin, bind DNA and block *Taq* polymerase activity. Others, e.g., EDTA or other chelating agents, act by sequestration of ions required for the enzymatic activity. Therefore, the amplification must be monitored using an internal control not related to the DNA target in AS-PCR and the absence of amplification in PCR-RFLP may indicate the presence of such substances (see Table 2.4).

2.2.9   Post PCR Analysis

PCR amplification performance and efficiency is routinely analysed by electrophoresis in agarose gel containing ethidium bromide. Amplified products are detected by visualization under ultraviolet (UV) light.

The gel concentration is dictated by the size of the amplified products or digested fragments. The resolving capability of a gel varies directly with its concentration: the higher the gel concentration, the greater will be the resolution. There are several types of agarose commercially available. Table 2.2 relates the type and concentration of agarose to the DNA fragment size to be resolved. Electrophoresis of DNA in agarose gel can be performed using either Tris-acetate EDTA (TAE) or Tris-borate EDTA (TBE).

Table 2.2

DNA base pair resolution for agarose in TAE and TBE

Agarose	% Agarose in TAE	% Agarose in TBE	Resolution
SeaKem	0.80	0.70	10,000–800 bp
	1.00	0.85	8,000–400 bp
	1.20	1.00	7,000–300 bp
	1.50	1.25	4,000–200 bp
	2.00	1.75	3,000–100 bp
SeaPlaque	0.75	0.70	25,000–500 bp
	1.00	0.85	20,000–300 bp
	1.25	1.00	12,000–200 bp
	1.50	1.25	6,000–150 bp
	1.75	1.50	3,000–100 bp
	2.00	1.75	2,000–50 bp
MetaPhor	2.00	1.80	800–150 bp
	3.00	2.00	600–100 bp
	4.00	3.00	250–50 bp
	5.00	4.00	130–20 bp
	–	5.00	< 80 bp

Electrophoresis in polyacrylamide gel is recommended for resolution of small differences between fragments of DNA. In some protocols, we recommend polyacrylamide gel for accuracy and clarity. The size of the fragments in the RFLP will dictate the gel type (agarose or polyacrylamide) and concentration to be used.

Table 2.3 gives acrylamide concentration, the respective resolution power for separating DNA fragment sizes, and the co-migration of the dyes present in the loading buffer (Sambrook et al 1989).

Table 2.3

DNA base pair resolution in polyacrylamide

Acrylamide (%(w/v))*	Fragment size resolution range in bp	Xylene cyanol FF†	Bromophenol blue†
3.5	1,000–2,000	460	100
5.0	80–500	260	65
8.0	60–400	160	45
12.0	40–200	70	20
15.0	25–150	60	15
20.0	6–100	45	12

* N,N'-methylenebisacrylamide:acrylamide at a concentration of 1:29.
† numbers indicate the approximate size of double-stranded DNA in bp that co-migrate with the dye.

2.2.10   Restriction Enzymes

Restriction enzymes are endonucleases, that is, enzymes that digest nucleic acids. Restriction enzymes recognize specific sequences of nucleotides in a DNA strand. Their use allows the detection of point mutations in DNA and eliminates the need for subcloning and sequencing. For instance, the *Alu* I restriction enzyme recognizes and cleaves the DNA sequence AGCT. If a DNA segment containing the AGCT sequence is subjected to *Alu* I treatment, it will be cleaved into two fragments. These fragments can be visualized after electrophoresis in agarose or acrylamide gel. If any one of the four nucleotides is replaced (e.g., GGCT or any other combination) the enzyme will no longer recognize the site and will not digest the DNA. There are enzymes originating from different sources (therefore different enzymes) that cleave within the same target sequence. These enzymes are known as iso-schizomers.

2.2.11   Controls

Every assay must include control DNA samples representative of each allele and combinations of alleles being tested. These controls are used to monitor PCR performance, amplification efficiency and enzyme treatment in RFLP analysis. Each PCR assay must also include a blank (all reagents except for DNA, which is replaced with the same volume of water) to monitor for contamination.

When performing AS-PCR an internal amplification control is required. This internal control must be of a sequence unrelated to the target sequence to avoid competition and must always be amplified to validate the presence or absence of an allele-specific reaction. PCR-RFLP does not require an unrelated internal control since every tube will have a product regardless of the allele.

2.3   Troubleshooting

Several factors can interfere with the successful outcome of performing molecular techniques (Table 2.4). Frequently, the user's manual provided with kits has a section with troubleshooting guides. Information and troubleshooting guides are also provided in the enzyme and reagent catalogues.

Table 2.4

Troubleshooting checklist

Problem	Possible cause	Possible solutions
No PCR product	Missing DNA or another reagent	Use a check mark system after adding each reagent
		Check all calculations to see if reagent concentrations are correct
	dNTP questionable quality	Take a fresh aliquot
	DNA quality	Perform electrophoresis on agarose gel to detect degradation
	DNA concentration too low	Use more DNA
	DNA concentration too high	Use less DNA
	Wrong $MgCl_2$ concentration	Optimize $MgCl_2$ titre for the specific reaction
	Wrong primer concentration	Optimize primer concentration for the specific reaction
	Primers degraded	Repeat reaction using new aliquot or check primers on a denaturing gel

	Primers with traces of ammonium	Ask manufacturer to check
	Inhibitory substances (poor DNA purity)	Perform PCR using half volume of DNA from the sample in question and half volume of DNA from a known positive sample. If no band is observed after amplification it is likely that the DNA sample contains inhibitory substances. Re-extract DNA and re-test
	Cycling conditions not optimal	Optimize cycling temperature and time for the specific reaction
	Poor ethidium bromide staining (ladder not seen also)	Stain with ethidium bromide for 5 minutes
	Excessive ethidium bromide staining (ladder seen)	Destain the gel in water for 30 to 60 minutes and analyse it again
	Hot start may be necessary	Try AmpliWax or HotStarTaq Polymerase
	Cycle profile was altered	Check to see file conditions unaltered
	Primer design not optimal	Review primer design. Check to see if primers are binding elsewhere
Smeary or multi-banded product	Wrong reagent concentration	Optimize reagents
	High concentration of DNA	Decrease DNA concentration
	Thermal cycling conditions not optimal	Optimize cycling profile or reduce number of cycles
	Primer design not optimal	Review primer design. Check to see if primers are binding elsewhere
	Carryover contamination	Observe for carryover prevention as in section 2.2.7
PCR contamination	Reagent contamination	Repeat PCR with brand new aliquot, if negative control clean and reaction still suspected of contamination, use newly extracted DNA sample
	Pre-PCR area contaminated	Clean up pre-PCR area and all equipment using 10% bleach and restart amplifying a slightly larger PCR product with at least one new primer
No digestion (samples and controls)	Restriction cleavage site is not present	Check that DNA sequence for site is present in amplified region
	Enzyme inactivated	Use a different source of enzyme Check expiration date
Incomplete digestion	Too much product used	Use half or less of product
	Too little enzyme used	Use 1.5×–2.0× more enzyme
	BSA needed	Check enzyme product monograph
	Incorrect incubation temperature	See enzyme requirement
	Incorrect reaction buffer	See enzyme requirement
Additional, atypical bands	Contaminated restriction enzymes	Perform enzyme treatment using a known sample Test sample with a new batch
	Altered site (new allele)	Test questionable DNA with another enzyme of known digestion pattern

2.4 References

Bianchi, D.W., Zickwolf, G.K., Weil, G.J., et al (1996) Male fetal progenitor cells persist in maternal blood for as long as 27 years postpartum. *Proc Natl Acad Sci USA* 93, 705–708.

Cha, R.S. and Thilly, W.G. (1993) Specificity, efficiency, and fidelity of PCR. *PCR Methods Appl* 3, S18–S29.

Cimino, G.D., Metchette, K.C., Tessman, J.W., et al (1991) Post-PCR sterilization: A method to control carryover contamination for the polymerase chain reaction. *Nucleic Acids Res* 19, 99–108.

Dieffenbach, C.W., Lowe, T.M.J. and Dveksler, G.S. (1993) General concepts for PCR primer design. *PCR Methods Appl* 3, S30–S37.

Dragon, E.A. (1993) Handling reagents in the PCR laboratory. *PCR Methods Appl* 3, S8–S9.

Fahy, E., Biery, M., Goulden, M., et al (1994) Issues of variability, carryover contamination, and detection in 2SR-based assays. *PCR Methods Appl* 3, S83–S94.

Gelfand, D.H. and White, T.J. (1990). Thermostable DNA polymerases. In *PCR Protocols: A Guide to Methods and Applications* (eds M.A. Innis, D.H. Gelfand, J.J. Sninsky et al), pp. 129–141. Academic Press, San Diego, CA.

Kwok, S. and Higuchi, R. (1989) Avoiding false positives with PCR. *Nature* 339, 237–238.

Mullis, K.B. (1991) The polymerase chain reaction in an anemic mode: how to avoid cold oligodeoxyribonuclear fusion. *PCR Methods Appl* 1, 1–4.

Prince, A.M. and Andrus, L. (1992) PCR: How to kill unwanted DNA. *BioTechniques* 12, 358–359.

Reed, W., Lee, T.-H., Busch, M.P., et al (1997). Sample suitability for the detection of minor leukocyte populations by polymerase chain reaction (PCR). Abstract. *Transfusion* 37 (Suppl), 107S.

Reid, M.E., Rios, M., Powell, V.I., et al (1999) DNA from blood samples can be used to genotype patients who have been recently transfused. *Transfusion*, in press.

Rios, M., Cash, K., Strupp, A., et al (1999) DNA from urine sediment or buccal cells can be used for blood group molecular genotyping. *Immunohematology*, 15, 61–65.

Sambrook, J., Fritsch, E.F. and Maniatis, T. (1989) Molecular Cloning: A Laboratory Manual (3 volumes), 2nd edition, Cold Spring Harbor Laboratory Press, Cold Spring Harbor, NY.

Wenk, R.E. and Chiafari, F.A. (1997) DNA typing of recipient blood after massive transfusion. *Transfusion* 37, 1108–1110.

Part 3   Protocols

3.1   Sample Shipment and Storage

Amniotic fluid for analysis and cell culture should be shipped in a sterile container at room temperature (not above 37°C). Short-term cultures can be established if the sample is <36 hours old. Samples for DNA isolation can be shipped frozen, at 4°C, or at room temperature. Isolated DNA is stored at 4°C, or at –20°C for long-term storage.

3.2   Reagents

The source and catalogue number for some common molecular biology reagents are given. Rarely are these the only source. They are given as a guide and are not an endorsement of a particular supplier.

3.3   Harvesting WBCs by RBC Sedimentation

Principle

When dextran is mixed with anticoagulated whole blood, the red cells form rouleaux and will separate from the WBCs and platelets by sedimentation.

Specimens

5 mL of anticoagulated blood (EDTA, citrate or ACD <2 days old; heparin <1 day old).

Equipment and Materials

Microcentrifuge
Micropipettors
Electronic serological pipettor
Filter barrier micropipette tips
15 mL conical polystyrene tubes
1.5 mL microcentrifuge tubes
Saline
Dextran T500 (Pharmacia, Cat #17-0320-01)
Sodium azide (Sigma Chemical Co., Cat #S2002)

Reagent Preparation

6% Dextran T500 in saline

To 75 mL of saline add:

Dextran T500   6.0 g
Sodium azide   0.2 g

When dissolved, make up to 100 mL with saline. Test the reagent prior to use as described in Quality Control (below).

Procedure

1. Combine 5 mL of whole blood, 5 mL of saline and 2 mL of 6% Dextran T500 in a 15 mL conical tube. Mix by inversion and let stand on the bench for ~1 hour.
2. Transfer the upper cell suspension (RBC poor) to another 15 mL tube.
3. Centrifuge the suspension at 160 × g for 10 minutes. Discard supernatant and resuspend the cells with 500–800 µL of saline.

Quality Control

To ensure that the dextran is isotonic, add one drop of fresh whole blood to 2 mL of 6% Dextran T500, mix and incubate at room temperature for 10 minutes. Centrifuge at $1,000 \times g$ and visually examine the supernatant for haemolysis.

Procedural Notes

1. Each millilitre of the RBC poor upper layer contains $3–5 \times 10^6$/mL WBCs.
2. Reticulocytes and aged RBCs do not sediment well since they do not form rouleaux.

Source

Ellis et al (1975); Tellez and Rubinstein (1970).

3.4   Short-term Cell Cultures from Amniotic Fluid

Principle

To cultivate and grow cells from amniotic fluid, which takes ~2 weeks using selective media.

Specimens

Sterile amniotic fluid (2.5 mL) in a sterile tube (<36 hours old).

Equipment and Materials

Laminar flow hood
Tissue culture incubator
Inverted microscope
5 mL, 10 mL serological pipettes
Electronic serological pipettor
T25 flask
Alcohol swabs
AmnioMax-C100 Basal Medium (Gibco BRL, Cat #17001)
AmnioMax-C100 Supplement (Gibco BRL, Cat #17002)

Procedure

1. Transfer the contents of the supplement to the basal medium using aseptic technique in a laminar flow hood. Mix and transfer 8 mL of complete sterile medium to a T25 flask labelled with the patient's name and the date. Securely cap the remaining culture medium and store at 2–8°C.
2. Transfer 2.5 mL of amniotic fluid to the T25 flask using aseptic technique. Loosely cap and then wipe the flask with an alcohol swab. Incubate the flask at 37°C containing 5% CO_2.
3. The next day, remove all but the last millilitre of medium and then add 9 mL of fresh medium using aseptic technique. Use an inverted microscope to monitor cell growth and for contamination. Wipe the flask with an alcohol swab and return to the incubator.
4. Change the medium every other day for the remainder of the culture time. Monitor cell growth for contamination; small foci of cells should be visible within 4 or 5 days.
5. A semi-confluent to confluent layer of cells should be obtained within 2 weeks. Use EDTA/trypsin to harvest the cells (Protocol 3.5).

Quality Control

To ensure that the culture medium is sterile, add 10 mL of medium to a T25 flask and incubate for at least 24 hours prior to using the medium.

Source

Epstein (1982); Hecht et al (1981).

3.5   Cultured Amniocyte Cell Harvesting

Principle

Cultured amniocytes are harvested from confluent layers grown in T25 flasks at 37°C in 5% CO_2. The cultured cells are washed with sterile PBS and then the adherent cells are detached from the flask by trypsin-EDTA digestion.

Specimens

One T25 flask containing a confluent layer of cultured cells from amniotic fluid (Protocol 3.4).

Equipment and Materials

Microcentrifuge
Micropipettors
Filter barrier micropipette tips
1.5 mL microcentrifuge tubes
0.45 μm filter
10 × Trypsin-EDTA (Gibco BRL, Cat #15400-096)
PBS, pH 7.4 (Gibco BRL, Cat #10010-023)
Fetal Bovine Serum (Gibco BRL, Cat #16000-036)
RPMI 1640 (Gibco BRL, Cat #11875-101)

Reagent Preparation

5 × Trypsin-EDTA
   Dilute Trypsin-EDTA 1:1 (v:v) with sterile PBS, pH 7.4.
   Aliquot into 1.5 mL microcentrifuge tubes and store at –20°C for no longer than 6 months.

10% Fetal Bovine Serum (FBS) in RPMI 1640
  Add 10 mL of FBS to 90 mL of RPMI 1640.
  Sterilize by filtration and store at 4°C for up to 3 months.

Procedure

1. Wash the adherent cells four times by filling the flask with PBS and decanting.
2. Invert the flask to drain on disposable tissue or blotting paper.
3. Add 1.5 mL of 5 × Trypsin-EDTA and incubate at 37°C for 10–15 minutes.
4. Gently tap the flask to loosen the adherent cells. Transfer the cells to a 1.5 mL microcentrifuge tube.
5. Centrifuge at 300 × g for 1 minute. Remove and discard the supernatant. Add 1.5 mL of 10% FBS in RPMI 1640. Resuspend the cells by vortex and centrifuge at 300 × g for 1 minute.
6. Remove and discard the supernatant. Resuspend the cells in PBS to ~2.5×10^7/mL (5×10^6 cells/200 µL).

3.6   Cell Preparation and DNA Extraction

Principle

To obtain DNA from eukaryotic cells using a chaotrophic reagent and proteinase K followed by precipitation with isopropanol or ethanol.

Specimens

Blood
Urine
Buccal/cervical smear
Amniotic fluid
Seminal fluid

Equipment and Materials

Microcentrifuge
Micropipettors
Filter barrier micropipette tips
1.5 mL microcentrifuge tubes
Absolute ethanol
Sterile saline
DNA isolation kit, e.g.: Easy DNA™ Kit (Invitrogen Corp., Cat #K1800-01) QIAamp DNA Mini Kit (Qiagen Inc., Cat #51104)

Reagent Preparation

Follow manufacturer's instructions.

Procedure

A.   Anticoagulated whole blood

1. Transfer 200 μL of blood to a microcentrifuge tube and extract DNA.

B.   Clotted blood

1. Mash the clot and transfer the cells to a 12 × 75 mm tube and centrifuge at $1,000 \times g$ for 5 minutes.
2. Remove the supernatant and transfer 200 μL of the interface layer to a 1.5 mL microcentrifuge tube and extract DNA.

C.   Frozen blood droplets

1. Place ~250 μL of frozen blood droplets into a microcentrifuge tube and allow to thaw.
2. Centrifuge at $1,000 \times g$ for 5 minutes, discard the supernatant, resuspend the cells in 200 μL of saline, and extract DNA.

D.   Donor unit segments

1. Transfer blood from the plastic tubing to a 1.5 mL microcentrifuge tube and centrifuge at $1,000 \times g$ for 5 minutes.
2. Remove the supernatant and transfer 200 µL of the interface layer to a 1.5 mL microcentrifuge tube and extract DNA.

E.   Urine

1. Centrifuge 15 mL urine at $1,000 \times g$ for 5 minutes.
2. Remove the supernatant, resuspend the pellet in 200 µL of saline and extract DNA.

F.   Buccal/cervical smear (cotton swab)

1. Remove the external layer of the cotton swab from the applicator stick and place into a 1.5 mL microcentrifuge tube containing 200 µL of saline.
2. Vortex briefly and extract DNA (note: the cotton will not interfere with DNA extraction).

G.   Amniotic fluid

1. Centrifuge 5–7 mL of amniotic fluid at $1,000 \times g$ for 5 minutes.
2. Remove the supernatant, resuspend the cells in 200 µL of saline and extract DNA.

H.   Seminal fluid

1. Add 1 mL of saline to 0.5 mL of seminal fluid and centrifuge at $1,000 \times g$ for 5 minutes.
2. Remove the supernatant, resuspend the cells in 200 µL of saline and extract DNA.

Quality Control

DNA quality should be assessed and the quantity determined (see Protocol 3.7).

3.7   Assessment of DNA Quality, Quantity and Purity

Principle

DNA quality can be analysed by electrophoresis in an agarose gel containing ethidium bromide. The DNA concentration is determined by optical density (OD) measurement at 260 nm. Alternatively, and when small amounts of DNA are extracted, the amount can be semi-quantitatively determined by comparison to DNA standards by agarose gel electrophoresis. Purity can be estimated by determining the DNA/protein ratio (OD 260/280 nm).

Specimen

Extracted DNA (Protocol 3.6).

Equipment and Materials

UV spectrophotometer
Micropipettors
Filter barrier micropipette tips
Agarose gel (see Protocol 3.11)
DNA standards
6 × DNA loading buffer (see Protocol 3.11)

Procedure

A.   Quality and semi-quantitation (agarose gel electrophoresis)

1. Combine 5 μL of DNA with 1 μL of 6 × DNA loading buffer and add to a lane of a 1% TAE agarose gel (Protocol 3.11). Include DNA standards.
2. Apply 100 volts of current for 30 minutes. Visualize the DNA under UV light. Photograph if desired.

3. Estimate the amount of DNA by visual comparison with the standards. High quality DNA appears as a relatively large single band.

B. DNA quantitation and purity (spectrophotometric analysis)

1. Determine the OD260 nm and OD280 nm (dilute the DNA in ddH_2O if necessary).
2. Calculate the amount of DNA using the following equation:
   DNA μg/mL = OD260 nm × dilution × 50 (DNA extinction coefficient).
3. Assess the purity by determining the OD260/280 ratio (see below).

Quality Control

A DNA standard should be measured to ensure proper calibration of the spectrophotometer.

Procedural Notes

1. The high molecular weight DNA (>20 kbp) is the fraction that can be efficiently amplified by *Taq* polymerase. Samples containing an excessive amount of degraded DNA may not be suitable for all PCR-based techniques.
2. A pooled DNA sample (frozen in 250 μL aliquots at –20°C) is a suitable 'in-house' standard for spectrophotometric readings. The value of the standard is recorded on a quality control log chart that provides the operator with acceptable limits of day-to-day variation (mean ± 2 SD).
3. DNA purity can be assessed by plotting the OD values from 230–400 nm (scanning spectrophotometer). Pure DNA has a typical sigmoid curve with a maximum OD at 260 nm. DNA contaminated with protein will show a 'shoulder' at 280 nm (the extinction coefficient for protein). DNA that is free from contaminating protein gives a ratio of 1.9 ± 0.1.

Comments

Specifically designed spectrophotometers are available with cuvettes that use <10 µL of sample (GeneQuant Calculator Cat #80-2109-98, Amersham Pharmacia Biotech). These instruments can detect as little as 35 ng of DNA.

3.8   PCR Amplification

Principle

Oligonucleotide primers, flanking a specific sequence of interest, are extended by *Taq* DNA polymerase (~1,000 nucleotides per second at 72°C) in the presence of deoxynucleotides and adequate salt solution. The primer annealing temperature and salt concentration are optimized to ensure gene-specific amplification. The amplification process occurs in rounds of three steps: (1) denaturation by heating the reaction to 94°C to 'melt' the DNA; (2) annealing of primers to complementary sequences; (3) polymerase-dependent elongation. After each round (or cycle) of denaturation, annealing and extension, the newly formed strands serve as template for the next cycle. It is estimated that a single DNA template will generate over a billion copies at the end of 30 cycles.

Specimens

Extracted DNA (Protocol 3.6).

Equipment and Materials

Microcentrifuge
Thermal cycler
Micropipettors (general use)
Micropipettors (PCR use only)

Filter barrier micropipette tips
0.5 and 0.2 mL microcentrifuge tubes
100 mM dNTP
10 × PCR buffer
25 mM $MgCl_2$
100 ng/µL primers
HotStarTaq Polymerase (Qiagen Inc., Cat. #203205) or *Taq* polymerase

Procedure

1. Select the primers that are appropriate for the polymorphism under analysis (see specific Protocols in Parts 4–6).
2. Prepare the appropriate cocktail (see PCR Cocktails below) for the total number of samples to be analysed. Make up a volume that is ~10% greater than the total volume required (include the samples, controls, and a water blank).
3. Add the appropriate amount of DNA (see PCR Cocktails below), vortex and flash spin tubes.
4. Amplify according to the appropriate profiles (see PCR Profiles below).
5. Combine 10 µL of the PCR product with 2 µL of 6 × loading buffer (Protocol 3.11) and electrophorese using an appropriate gel system (Protocols 3.11 and 3.12).
6. Photograph the gel. The expected PCR product should be clearly visible. If no bands are visible, or there are unexpected bands, refer to the troubleshooting checklist in Part 2 (Table 2.4).

PCR Cocktail A

Reagent	One sample	_____ samples
10 × buffer II	5.0 µL	_____ µL
25 mM $MgCl_2$	5.0 µL	_____ µL
1.25 mM dNTPs	4.2 µL	_____ µL
Sense primer (100 ng/µL)	0.7 µL	_____ µL
Antisense primer (100 ng/µL)	0.7 µL	_____ µL
Taq polymerase (5.0 U/µL)	0.25 µL	_____ µL
Sterile ddH_2O	24.15 µL	_____ µL
Total	40.0 µL	_____ µL

Combine 40 µL of the cocktail with 10 µL of DNA (50–200 ng total) per tube.

PCR Cocktail B

Reagent	One sample	_____ samples
10 × buffer II	5.0 µL	_____ µL
20 mM MgCl$_2$	5.0 µL	_____ µL
1.25 mM dNTPs	4.2 µL	_____ µL
Sense primer (100 ng/µL)	0.5 µL	_____ µL
Antisense primer (100 ng/µL)	0.5 µL	_____ µL
Taq polymerase (5.0 U/µL)	0.25 µL	_____ µL
Sterile ddH$_2$O	24.55 µL	_____ µL
Total	40.0 µL	_____ µL

Combine 40 µL of the cocktail with 10 µL of DNA (100–200 ng total) per tube.

PCR Cocktail C

Reagent	One sample	_____ samples
10 × buffer with 15 mM MgCl$_2$	5.0 µL	_____ µL
10 mM dNTPs	1.0 µL	_____ µL
Sense primer (100 ng/µL)	1.0 µL	_____ µL
Antisense primer (100 ng/µL)	1.0 µL	_____ µL
HotStarTaq polymerase (5.0 U/µL)	0.3 µL	_____ µL
Sterile ddH$_2$O	31.7 µL	_____ µL
Total	40.0 µL	_____ µL

Combine 40 µL of the cocktail with 10 µL of DNA (100–200 ng total) per tube.

PCR Cocktail D

Reagent	One sample	_____ samples
10 × buffer II	5.0 µL	_____ µL
25 mM MgCl$_2$	3.0 µL	_____ µL
2 mM dNTPs	5.0 µL	_____ µL
Sense primer (100 ng/µL)	1.0 µL	_____ µL
Antisense primer (100 ng/µL)	1.0 µL	_____ µL
HotStarTaq polymerase (5.0 U/µL)	0.4 µL	_____ µL
Sterile ddH$_2$O	14.6 µL	_____ µL
Total	30.0 µL	_____ µL

Combine 30 µL of the cocktail with 20 µL of DNA (500 ng total) per tube.

PCR Profile 1

94°C for 5 or 15 minutes, respectively, for *Taq* polymerase and HotStarTaq.
94°C for 20 seconds, 62°C for 20 seconds, 72°C for 20 seconds for 35 cycles.
72°C for 10 minutes, and 4°C hold.

PCR Profile 2

94°C for 5 or 15 minutes, respectively, for *Taq* polymerase and HotStarTaq.
94°C for 20 seconds, 60°C for 20 seconds, 72°C for 20 seconds for 35 cycles.
72°C for 10 minutes, and 4°C hold.

PCR Profile 3

94°C for 5 or 15 minutes, respectively, for *Taq* polymerase and HotStarTaq.
94°C for 30 seconds, 62°C for 60 seconds, 72°C for 60 seconds for 34 cycles.
72°C for 10 minutes, and 4°C hold.

PCR Profile 4

94°C for 5 or 15 minutes, respectively, for *Taq* polymerase and HotStarTaq.
94°C for 20 seconds, 64°C for 15 seconds, 72°C for 30 seconds for 5 cycles.
94°C for 20 seconds, 64°C for 15 seconds, 72°C for 15 seconds for 25 cycles.
72°C for 10 minutes, and 4°C hold.

Procedural Notes

The following precautions will help avoid contamination (for more details see Part 2).

1. Wear a separate laboratory coat in each area.
2. Change gloves frequently during PCR set-up and when moving between analytical areas.
3. Combine the PCR reagent cocktails before handling the DNA samples.
4. Use positive displacement pipettors committed for PCR reagents only.
5. Use filter barrier micropipette tips throughout the procedure.
6. Wipe equipment (racks, heat blocks, and centrifuge bowls) with bleach diluted 1:10 in water. Equipment used to isolate DNA should be committed for that use only.
7. Keep the workbench clean. It is recommended to change bench padding daily and wipe the workbench area with dilute bleach once a week or if spills occur.

Quality Control

1. Contamination is detected by testing a water blank reagent control in place of DNA.
2. DNA from individuals who are known homozygous and heterozygous for the polymorphism of interest is analysed each time.

Results

1. The water blank must not have any visible amplified product.
2. Test and control samples must have a single amplified product of the expected size.

Comments

Generally, 100–500 ng of DNA ($3–16 \times 10^4$ copies for autosomal copy genes) will yield 0.5–1 µg of amplified product.

3.9   Restriction Enzyme Digestion for PCR-RFLP Analysis

Principle

Digestion of a PCR amplified product yields fragments of predicted size based on the number and position of the restriction endonuclease cleavage sites.

Specimens

Amplified DNA (Protocol 3.8).

Equipment and Materials

Heat block
Micropipettors (general use)
Restriction enzyme

Procedure

1. Prepare a restriction enzyme mix for the number of samples to be analysed plus 1 ($n + 1$) for every 10 tests.

Reagent	One sample	_____ samples
10 × buffer	2.0 µL	_____ µL
Restriction enzyme	___ µL (to have 5U)	_____ µL
Sterile ddH$_2$O	___ µL(to bring to 12 µL)	_____ µL

Note: Some enzymes require BSA. When using BSA, adjust the volume of water accordingly.

2. Combine 12 µL of the mix to a 0.5 mL microfuge tube with 8 µL of DNA. Mix and flash spin.
3. Incubate at the temperature given by the manufacturer for a minimum of 3 hours.

4. After incubation, add 4 µL of 6 × DNA loading buffer (Protocol 3.11) to the tube, vortex and flash spin.
5. Perform agarose gel electrophoresis (Protocol 3.11) using an appropriate concentration of agarose (see Part 2, Table 2.2).

Procedural Notes

1. The manufacturer of the restriction enzyme provides the appropriate diluent for optimum reactivity.
2. One restriction enzyme unit (U) is defined as the amount of enzyme required to digest 1 µg of DNA in 1 hour at the optimal temperature. To compensate for enzyme degradation and small differences in salt concentrations due to PCR buffers, it is recommended to use 5 U of enzyme and to incubate for at least 3 hours to ensure complete digestion.
3. Storage buffer for restriction enzymes contains glycerol, which can inhibit the enzyme activity if its final concentration exceeds 5%.

Limitations of the Procedure

1. There is the potential for error using PCR-RFLP when an allele does not have a restriction endonuclease site. To ensure complete enzymatic digestion, always include known homozygous and heterozygous samples for analysis.

3.10   Allele-specific PCR

Principle

To detect single nucleotide polymorphisms using sequence-specific primers by PCR. This approach is used when neither polymorphism is associated with a restriction enzyme site.

Specimens

Extracted DNA (Protocol 3.6).

Equipment and Materials

Microcentrifuge
Thermal cycler
Micropipettors (general use)
Micropipettors (PCR use only)
Filter barrier micropipette tips
1.5 mL, 0.5 mL, 0.2 mL microcentrifuge tubes
$10 \times$ PCR buffer
25 mM $MgCl_2$
dNTPs (2 mM)
Nuclease-free ddH_2O
Allele-specific primers (100 ng/µL)
Internal control sense/antisense primers (100 ng/µL)
HotStarTaq Polymerase (Qiagen Inc., Valencia, CA, Cat #203205)

Procedure

1. Select the sequence-specific primers that are appropriate for the polymorphism under analysis (see specific Protocols in Parts 4–6).
2. For each allele, prepare a cocktail (see AS-PCR Cocktail below) for the total number of samples to be analysed. Make up a volume that is 10% greater than the total volume required (include the samples, controls and a water blank (e.g. $n + 1$)).
3. Add an appropriate amount of DNA (see AS-PCR Cocktail below), vortex and flash spin tubes.
4. Amplify according to the AS-PCR Profile below.
5. Combine 10 µL of the PCR product with 2 µL of $6 \times$ loading buffer (Protocol 3.11) and electrophorese using an appropriate gel system (Protocols 3.11 and 3.12).
6. Photograph the gel. If unexpected bands are found, refer to the troubleshooting checklist in Part 2 (Table 2.4).

AS-PCR Cocktail

Reagent	One sample	_____ samples
10 × buffer	5.0 µL	_____ µL
25 mM MgCl$_2$	3.0 µL	_____ µL
2 mM dNTPs	5.0 µL	_____ µL
Allele 1 or 2 sequence-specific primer (100 ng/µL)	1.0 µL	_____ µL
Common antisense primer (100 ng/µL)	1.0 µL	_____ µL
Internal control sense primer (100 ng/µL)	0.5 µL	_____ µL
Internal control antisense primer (100 ng/µL)	0.5 µL	_____ µL
HotStarTaq polymerase (5.0 U/µL)	0.25 µL	_____ µL
Sterile ddH$_2$O	23.75 µL	_____ µL
Total	40.0 µL	_____ µL

Combine 40 µL of cocktail for each allele with 10 µL of DNA (50–200 ng total) in a tube.

AS-PCR Profile 1

94°C for 15 minutes for HotStarTaq polymerase.
94°C for 20 seconds, 55°C for 20 seconds, 72°C for 20 seconds for 30 cycles.
72°C for 10 minutes, and 4°C hold.

AS-PCR Profile 2

94°C for 15 minutes for HotStarTaq polymerase.
94°C for 20 seconds, 57°C for 20 seconds, 72°C for 20 second for 30 cycles.
72°C for 10 minutes, and 4°C hold.

Procedural Notes

1. The precautions listed under Procedural Notes in Protocol 3.8 (PCR Amplification) will help avoid contamination. For additional considerations see Part 2.

2. The addition of *Taq* polymerase directly to the reaction mix along with DNA is possible for certain types of *Taq* polymerases. For example, the enzymatic activity of HotStarTaq polymerase is inhibited at room temperature but is activated upon heating above the annealing temperature. Therefore, mis-priming is avoided.

Quality Control

1. The water blank ('reagent control') must not have any bands.
2. DNA from individuals who are known homozygous and heterozygous for the polymorphism of interest must be analysed each time.

Results

1. All tests and control samples must have the internal control band of the expected size.
2. The test sample should have a band if the polymorphism is present or no band in the absence of the polymorphism.

3.11   Agarose Gel Preparation and Electrophoresis

Principle

DNA fragments are separated by molecular sieving through agarose in a buffered-salt solution using a horizontal electrophoresis apparatus.

Specimens

Extracted DNA (Protocol 3.6).
PCR amplified products (Protocol 3.8).
Restriction enzyme digested PCR products (Protocol 3.9).

Equipment and Materials

Camera and polaroid black and white film (Propan 667)
Microwave oven
Gel electrophoresis apparatus
UV transluminator
Micropipettors
50 × TAE or 10 × TBE
6 × DNA loading buffer
Sterile ddH$_2$O
DNA ladder
10 mg/mL ethidium bromide
High resolution agarose, e.g.:
  NuSieve 3:1 Agarose (FMC BioProducts, Cat #500980)
  Agarose-1000 (Gibco BRL, Cat #10975-027)
Multi-purpose agarose, e.g.:
  Analytical Grade Agarose (Bio-Rad Laboratories, Cat #162-0125)
  UltraPure™ Agarose (Gibco BRL, Cat #15510-027)

Reagent Preparation

The reagents can be purchased and must be of molecular grade.

50 × TAE

Tris base	242 g
Glacial acetic acid	57.1 mL
0.5 M EDTA (pH 8.0)	100 mL

Dissolve completely to a final volume of 1 L in ddH$_2$O.

10 × TBE

Tris base	108 g
Boric acid	55 g
0.5 M EDTA (pH 8.0)	40 mL

Dissolve completely to a final volume of 1 L in ddH$_2$O

6 × DNA loading buffer

Ficoll 400	15 g
Xylene cyanol	0.25 g
Bromophenol blue	0.25 g
Sterile ddH$_2$O	100 mL

Aliquot in 1.5 mL microcentrifuge tubes.

Procedure

1. Weigh an appropriate amount of agarose into a flask and add electrophoresis buffer (1 × TAE or 0.5 × TBE) sufficient to obtain the desired concentration.
2. Cover the top of the flask with DuraSeal™ or similar covering. Melt the agarose in a microwave oven.
3. When the gel has cooled to ~55°C (hand-hot), add ethidium bromide (0.5 μg/mL final), mix by gentle swirling.
4. Depending on the apparatus, seal the ends of a gel platform with tape or place the platform in a casting tray.
5. Pour the warm agarose into the platform and position the comb in the gel. Allow the gel to set for at least 1 hour.
6. Remove the comb, taking care not to tear the wells. If necessary, remove the tape from the ends of the gel platform.
7. Add to the chamber, sufficient buffer to cover the gel. Ensure that no air pockets are trapped within the wells.
8. Prepare samples for running by mixing the appropriate amount of sample and 6 × DNA loading buffer as indicated:
   (a) DNA sample: Combine 5 μL of DNA with 1 μL of 6 × DNA loading buffer and transfer the entire contents into a well.
   (b) PCR product: Combine 10 μL of amplified product with 2 μL of 6 × DNA loading buffer and transfer the entire contents into a well.
   (c) Digestion product: Combine 3 μL of 6 × DNA loading buffer to 15 μL of the digestion reaction and transfer the entire contents into a well.
9. Add 10 μl of the appropriate DNA molecular size marker into a well.
10. Attach the electrode leads such that the DNA will migrate towards

the anode. Electrophorese at 60–100 volts. The progress of separation can be monitored by the migration of the dyes in the gel.
11. Turn off the power supply when the bromophenol blue has migrated a distance judged sufficient for DNA resolution.
12. Photograph the gel using a UV light source.

Procedural Notes

1. Gels are typically 1 cm thick. However, it is important to ensure that the volume of the sample does not exceed the capacity of the well. The volume of the well is determined by the width of the comb and the gel height.
2. The agarose concentration and type (multi-purpose versus high resolution) will depend on the size of the DNA fragments and the desired resolution (see Part 2 for a general guideline).

Safety

1. Agarose can become 'superheated' when boiled using a microwave. Use heat-protective gloves when preparing molten agarose.
2. Ethidium bromide is a potential carcinogen. Wear gloves to avoid contact with skin.
3. Ultraviolet radiation is dangerous to the eyes and exposed skin. Ensure that the ultraviolet light source is shielded or wear a protective face-shield and clothing.
4. Observe precautions for working with high voltage.

3.12   Acrylamide Gel Preparation and Electrophoresis

Principle

Separation of small DNA fragments or DNA fragments of a similar size in a non-denaturing polyacrylamide gel.

Specimen

PCR amplified products (Protocol 3.8).
Restriction enzyme digested PCR products (Protocol 3.9).

Equipment and Materials

Camera and black and white film (Polaroid Propan 667)
Electrophoresis apparatus (Bio-Rad Laboratories, Hercules, CA)
UV transluminator
Micropipettors
10 × TBE (Protocol 3.11)
6 × DNA loading buffer (Protocol 3.11)
Sterile ddH$_2$O
10 mg/mL ethidium bromide
DNA ladder
Ammonium persulphate
TEMED
29:1 acrylamide/N,N'-bisacrylamide

Procedure

1. Assemble and clamp together two glass plates with spacers.
2. Prepare the polyacrylamide solution in a beaker by adding together:

	8%	6%
acrylamide/N,N'-bisacrylamide	23.3 mL	17.5 mL
10 × TBE	8.8 mL	8.8 mL
ddH$_2$O	55 mL	61 mL
10% ammonium persulphate	363 μL	363 μL
TEMED	66 μL	66 μL

3. Carefully pour the solution between the plates and insert a comb.
4. Allow the acrylamide to polymerize for at least 1 hour.
5. Remove the comb and bottom spacer and clamp the glass assembly into position.

6. Fill both upper and lower chambers with 1 × TBE buffer.
7. Load samples and the molecular size marker into the wells.
8. Electrophorese at 200 volts for 2 hours or until the bromophenol blue has migrated to within 2 cm of the bottom of the gel.
9. Remove the plate assembly and transfer the gel into a tray containing ~50 mL of 1 × TBE and 10 µL of 10 mg/mL ethidium bromide. Mix for 2 minutes.
10. Discard the solution and add ddH$_2$O to cover the gel. Allow the gel to destain for 5–10 minutes. Photograph gel using a UV light source.

Procedural Notes

See Table 2.3 for acrylamide concentrations and their ability to resolve DNA fragments.

Safety

1. Acrylamide/N,N'-bisacrylamide is a neurotoxin, and ethidium bromide is a potential carcinogen. Avoid skin contact with these reagents.
2. Ultraviolet radiation is dangerous to the eyes and exposed skin. Ensure that the ultraviolet light source is shielded or wear a protective face-shield and clothing.
3. Observe precautions for working with high voltage.

Source

Ausubel et al (1997); Sambrook et al (1989).

3.13   Single-strand Conformation Polymorphism (SSCP)

Principle

Single nucleotide substitutions alter the mobility of single-stranded DNA during polyacrylamide gel electrophoresis.

Specimens

PCR amplified DNA spanning a polymorphism.

Equipment and Materials

Microcentrifuge
Heat block
UV transluminator
Electrophoresis apparatus
Micropipettors (general use)
Microcentrifuge tubes
Filter barrier tips
Scalpel
Qiaex II Gel Extraction System (Qiagen, Cat #20021)
6% polyacrylamide gel system (Protocol 3.12)
1.2% agarose gel
Gelstar Reagent (FMC, Cat #50535)
Crushed ice
Loading buffer

Procedure

1. Electrophorese the PCR amplified DNA (Protocol 3.8) using a 1.2% agarose gel (Protocol 3.11).
2. Place the gel under UV light. Excise the band of interest from the gel using a scalpel. Transfer the band to a 1.5 mL microcentrifuge tube.
3. Purify the DNA from the band using Qiaex II according to manufacturer's instructions.
4. Semi-quantify the purified DNA (Protocol 3.7A).
   Combine 5 µl of the purified DNA with 5 µl of loading buffer and mix.
5. Heat the reaction mix at 100°C for 3 to 5 minutes and then place the tube immediately on crushed ice.
6. Transfer the sample to a lane of a non-denaturing 6% polyacrylamide gel.
7. Electrophorese for 17 h at 35 V.

8. Transfer the gel to a tray and stain with Gelstar solution.
9. Photograph the gel under UV light.

Quality Control

Include PCR amplified DNA from heterozygous and homozygous individuals.

Reagent Preparation

Loading buffer

95% Formamide
20 mM EDTA
0.05% Bromophenol blue
0.05% Xylene cyanol

Gelstar working solution

Dilute the reagent according to the manufacturer's instructions.

Reporting Results

1. The results of the test samples are based on a comparison of the banding patterns against control samples.
2. Test samples showing a banding pattern that does not match one of the controls cannot be assigned a genotype and should be subject to DNA sequencing.

Procedural Notes

The addition of 10% glycerol to the polyacrylamide gel during preparation will retard sample diffusion. This is particularly useful for fragments that are 100 bp or smaller.

Limitations of the Procedure

1. Mutations that do not produce changes to the secondary structure of the single-stranded DNA will not be detected by this procedure.
2. Some band shifts can be too subtle to be resolved using this technique.

Comments

1. DNA products >300 bp may require restriction enzyme digestion prior to SSCP analysis.
2. The temperature of the gel during electrophoresis may affect resolution.
3. Optimization of the assay is required for each DNA sequence to be analysed, i.e., gel concentration, temperature during electrophoresis, and voltage.
4. Pre-made polyacrylamide gels are commercially available for this particular procedure.
5. Observe safety precautions when working with acrylamide.

Source

Orita et al (1989a, 1989b).

3.14   Optimizing a PCR

Principle

The PCR can show non-specific amplification under less than ideal conditions. The process outlined in this protocol provides guidelines to vary the reagents, annealing temperature, and number of cycles to optimize the reaction.

Specimens

Extracted DNA (Protocol 3.6).

Equipment and Materials

Microcentrifuge
Thermal cycler
Micropipettors (general use)
Micropipettors (PCR use only)
Filter barrier micropipette tips
0.2 mL microcentrifuge tubes
100 mM dNTP
10 × PCR buffer
25 mM MgCl$_2$
Formamide
HotStarTaq Polymerase (Qiagen Inc., Cat #203205) or *Taq* polymerase

Procedure

A. Temperature and number of cycles optimization

1. Prepare the PCR cocktail as described below for several (10–12) test samples. Make up a volume that is ~10% greater than the total volume required (include the samples, controls and water blank).
2. Aliquot 30 µL of PCR cocktail into 0.2 µL PCR microcentrifuge tubes.
3. Add 150 ng, 250 ng or 500 ng of DNA (three representative samples of each), vortex and flash spin tubes.
4. Amplify according to the PCR profile below. The annealing temperature is the Tm and is based on the lower of the two amplification primers based on the following equation: Tm = 2 (A+T) + 4 (G+C).
5. The elongation temperature (72°C) is 60 seconds for fragments that are ~500 bp or less and 120 seconds for fragments that are >500 bp.
6. Combine 10 µL of the PCR product with 2 µL of 6 × loading buffer and electrophorese using an appropriate agarose gel system (Protocol 3.11).
7. Photograph the gel. The expected PCR products may be visible for one of the DNA concentrations. See Results section below (items 1–3) for additional considerations.

B. Magnesium ion optimization

1. Prepare four PCR cocktails as described below with a final magnesium concentration of 1.0 mM (2.0 µL), 1.5 mM (3.0 µL),

2.0 mM (4.0 μL), and 3.0 mM (5.0 μL). Reduce the amount of water in each cocktail accordingly.
2. Aliquot the cocktail and add DNA as described in section A.
3. Choose the PCR profile with the annealing temperature and number of cycles based on the optimum results thus far. See Results (item 4) for additional considerations.

C. Formamide optimization

1. Prepare six PCR cocktails as described below with a final formamide concentration of 0%, 0.5%, 1.0%, 2.5%, 4.0% and 5.0%. Reduce the amount of water in each cocktail accordingly.
2. Aliquot the cocktail and add DNA as described in section A.
3. Choose the PCR profile with the annealing temperature, number of cycles, and magnesium concentration based on the optimum results thus far.

PCR cocktail

Reagent	One sample	_____ samples
10 × buffer	5.0 μL	_____ μL
25 mM MgCl$_2$	3.0 μL	_____ μL
2 mM dNTPs	5.0 μL	_____ μL
Sense primer (2.5 nmol)	1.0 μL	_____ μL
Antisense primer (2.5 nmol)	1.0 μL	_____ μL
Formamide (optional)	(_____ μL)	(_____ μL)
HotStarTaq polymerase (5.0 U/μL)	0.4 μL	_____ μL
Sterile ddH$_2$O (adjust if necessary)	14.6 μL	_____ μL
Total	30.0 μL	_____ μL

PCR profile

94°C for 5 minutes for *Taq* polymerase or 15 minutes for HotStarTaq.
94°C for 30 seconds, Tm°C for 30 seconds, 72°C for 60 seconds (<500 bp) or 120 seconds (>500 bp) for 35 cycles.
72°C, 10 minutes, and 4°C hold.

Procedural Notes

See Protocol 3.8 for precautions to help avoid contamination.

Quality Control

Contamination is detected by testing a water blank reagent control in place of DNA.

Results

1. The water blank must not have any visible amplified product.
2. If no band is visible for all DNA concentrations, repeat the PCR at an annealing temperature that is 1°C or 2°C lower. If the expected fragment is the only band but is weakly visible for only one of the DNA concentrations, increase the number of PCR cycles by 2 for that concentration of DNA. If the expected fragment and non-specific bands are visible for all DNA concentrations, repeat the PCR at an annealing temperature that is 1°C or 2°C higher.
3. If raising the annealing temperature results in a decrease in the amplification of the specific fragment along with a similar decrease in the non-specific bands for all DNA concentrations, then go to section B, 'Magnesium ion optimization', above.
4. If varying the magnesium concentration results in decrease or increase in the amplification of the specific fragment along with a similar decrease or increase in non-specific bands for all DNA concentrations, then go to section C, 'Formamide optimization', above.

Comments

1. It is not recommended to amplify beyond 40 cycles.
2. If the addition of formamide does not result in the amplification of the specific fragment only, then consider redesigning the amplification primers.
3. Optimization kits are commercially available.

3.15   Gene Zygosity Assessment by Quantitative PCR Amplification

Principle

PCR amplification of DNA can be quantitatively measured for genes that are present in either the homozygous or hemizygous (absent on one of the autosomal chromosomes) state, e.g., *RHD*. Sequence specific primers used to amplify a region of a test gene are compared to the relative amount of PCR product amplified for a control gene in a multiplex PCR.

Specimens

Extracted DNA (Protocol 3.6).

Equipment and Materials

Microcentrifuge
Thermal cycler
Computing Densitometer and ImageQuant™ Software (Molecular Dynamics Inc., Sunnyvale, CA)
Micropipettors (general use)
Micropipettors (PCR use only)
Filter barrier micropipette tips
0.2 mL PCR tubes
100 mM dNTP
10 × PCR buffer
25 mM $MgCl_2$
10 mg/mL Bovine Serum Albumin (Amersham Pharmacia Biotech, Cat #27-8915-01)
100 ng/µL primers:
    pair of primers that amplify a non-polymorphic sequence on an autosomal chromosome (internal control for amplification)
    test primers for a homozygous/hemizygous gene system (e.g., *RHD*)
HotStarTaq Polymerase (Qiagen Inc., Cat #203205)

Procedure

A. Quantitative PCR

1. Prepare the PCR cocktail (see below) for the total number of samples to be analysed. Make up a volume that is ~10% greater than the total volume required (include the samples, controls, and a water blank).
2. Aliquot 30 µL of PCR cocktail into 0.2 µL PCR microcentrifuge tubes.
3. Add 125 ng of DNA, vortex and flash spin tubes.
4. Amplify according to the PCR profile below.
5. Combine 10 µL of the PCR product with 2 µL of 6 × loading buffer and electrophorese using a 3% agarose gel system (Protocol 3.11).
6. Photograph the gel using Kodak 665 film. The expected PCR products should be clearly visible. If no bands are visible, or there are unexpected bands, refer to the troubleshooting checklist in Part 2 (Table 2.4).
7. Immerse the negative portion of the film in a tray of tap water for 10–15 minutes.
8. Gently rub the film and rinse under running tap water to remove any residual developing reagent from the film. The film can be detached from the processing paper at this time.
9. Air-dry the negative film.

B. Densitometer analysis

1. Set background and scan the negative film according to the manufacturer's instructions.
2. Using a 'rectangular' object, highlight the area of the control band for the homozygous control sample. Determine and record the densitometric intensity for this band. Save this rectangle area for use on all other bands.
3. Move the rectangle to the test band for the homozygous control sample and determine the densitometric intensity.
4. Continue with densitometric scanning for the bands of the remaining test and control samples.
5. Calculate the test/control ratio (the test band intensity divided by the control band intensity) for each test and control sample.

PCR cocktail

Reagent	One sample	_____ samples
10 × buffer	5.0 µL	_____ µL
25 mM MgCl$_2$	3.0 µL	_____ µL
2 mM dNTPs	5.0 µL	_____ µL
Primers (100 ng/µL)	1.0 µL each	_____ µL
BSA 10 mg/mL	0.5 µL	_____ µL
HotStarTaq polymerase (5.0 U/µL)	0.4 µL	_____ µL
Sterile ddH$_2$O	12.1 µL	_____ µL
Total	30.0 µL	_____ µL

PCR profile

94°C for 15 minutes.
94°C for 30 seconds, 49°C for 30 seconds, 72°C for 60 seconds, for 40 cycles.
72°C for 10 minutes, and 4°C hold.

Procedural Notes

See Protocol 3.8 and Part 2 for precautions to help avoid contamination.

Quality Control

1. Contamination is detected by testing a water blank reagent control in place of DNA.
2. Known homozygous and hemizygous samples must be included with each assay. The test/control ratio for the controls must be ± 2 SD of the mean (see Comments below).

Results

1. The water blank must not have any visible amplified product.

2. The ratio for test samples must fall within either the homozygous or hemizygous control range.

Comments

1. A mean and range (± 2 SD) for the test/control ratio should be established using at least 30 homozygous and 30 hemizygous samples. There must be no overlap between the homozygous and hemizygous ranges. The mean values obtained from the homozygous and hemizygous controls may not necessarily reflect the actual copy numbers.
2. A quality control log sheet should be used to record the daily results for the controls. Periodic inspection of a graphic representation of the control values (ratio versus test date) reveals any drift or trends that occur.
3. The quantitative PCR assay is particularly sensitive to acid hydrolysis of the amplification primers. It is recommended that the stock primers be stored at –70°C. The primers should be replaced if quality control graphs indicate unacceptable variation to the ratio.

Source

Denomme and Fernandes (1999).

3.16   VNTR and STR Analysis by PCR Amplification

Principle

Variable number of tandem repeat (VNTR) and short tandem repeat (STR) sequences are inherited repeating stretches of DNA from hundreds (VNTRs) to a few (STRs) nucleotides generally found in the non-coding regions of the human genome. These stretches of DNA show size heterogeneity among humans and are inherited in a co-dominant fashion. In the test, oligonucleotide primers, flanking a VNTR/STR, are extended using the PCR. Agarose or polyacrylamide gel electrophoresis is used to estimate the PCR fragment sizes, respectively VNTR and STR.

Specimens

Extracted DNA (Protocol 3.6); maternal, paternal (optional) and amniotic fluid or cervical smear DNA.

Equipment and Materials

Microcentrifuge
Thermal cycler
Micropipettors (general use)
Micropipettors (PCR use only)
Filter barrier micropipette tips
0.2 mL PCR tubes
100 mM dNTP
10 × PCR buffer
25 mM $MgCl_2$
VNTR primers (Genome Database: http://www.hgmp.mrc.ac.uk/gdb/)
  D1S80-PCR1.1/PCR1.2 (GDB:178639)
    5'-GAAACTGGCCTCCAAACACTGCCCGCCG-3'
    5'-GTCTTGTTGGAGATGCACGTGCCCCTTGC-3'
  D17S5-YNZ22.1/YNZ22.2 (GDB:178624)
    5'-CACAGTCTTTATTCTTCAGCG-3'
    5'-CGAAGAGTGAAGTGCACAGG-3'
  ApoB-PCR6.1/PCR6.2 (GDB:177687)
    5'-GGACAGTGAAACGAGGGC-3'
    5'-GGCACATGAAGACACCAGAGG-3'
STR primers (Maxim Biotech, Inc., San Francisco, CA):
  HPRT (Cat #STR-1001/STR-1002)
  Tyrosine hydroxylase (Cat #STR-1003/STR-1004)
  Androgen receptor (Cat #STR-1007/STR-1008)
  Intestinal Fatty Acid Binding Protein (Cat #STR-1009/STR-10010)
HotStarTaq Polymerase (Qiagen Inc., Cat #203205) or *Taq* polymerase

Procedure

1. Prepare the appropriate cocktail (see PCR Cocktail below) for the total number of samples to be analysed. Make up a volume that is ~10% greater than the total volume required (include the samples, controls and a water blank).
2. Aliquot 30 µL of PCR cocktail into 0.2 µL PCR tubes.
3. Add 250 ng of DNA, vortex and flash spin tubes.
4. Amplify according to the PCR profile below.
5. Combine 10 µL of the PCR product with 2 µL of 6 × loading buffer and electrophorese using 3.0% NuSieve agarose (for VNTR, see Protocol 3.11) or 10% polyacrylamide gel (for STR, see Protocol 3.12).
6. Photograph the gel. The expected PCR products should be clearly visible. If no bands are visible, or there are unexpected bands refer to the troubleshooting checklist in Part 2 (Table 2.4).

PCR cocktail

Reagent	One sample	_____ samples
10 × buffer	5.0 µL	_____ µL
25 mM $MgCl_2$	3.0 µL	_____ µL
2 mM dNTPs	5.0 µL	_____ µL
Sense primer (2.5 nmol)	1.0 µL	_____ µL
Antisense primer (2.5 nmol)	1.0 µL	_____ µL
HotStarTaq polymerase (5.0 U/µL)	0.4 µL	_____ µL
Sterile ddH_2O	14.6 µL	_____ µL
Total	30.0 µL	_____ µL

PCR Profile

94°C for 5 minutes for *Taq* polymerase, or 15 minutes for HotStarTaq.
94°C for 30 seconds, 60°C for 60 seconds, 72°C for 120 seconds for 35 cycles.
72°C for 10 minutes, and 4°C hold.

Procedural Notes

See Protocol 3.8 and Part 2 for precautions to help avoid contamination.

Quality Control

1. Contamination is indicated by testing a water blank reagent control in place of DNA.
2. Paternal DNA can be tested to confirm VNTR/STR inheritance patterns when homozygous fragments (single band) are detected in the fetal-derived DNA.
3. When VNTR/STR analysis detects maternal DNA in the fetally-derived sample, the antigen status of the potentially compatible fetus is not confirmed unless the blood group genotype has been validated for a similar range of DNA mixtures. Alternatively, amniocytes can be cultured (Protocols 3.4 and 3.5) to confirm the fetal blood group antigen status.

Results

1. The water blank must not have any visible amplified product.
2. Test and control samples will have no more than two amplified products of the approximate size as shown below.

Tandem repeat	Nucleotide repeat	Repeat size (bp)
Apo B	30	825
D1S80	16	530
D17S5	70	378
HR-1001/2	4	283
HR-1003/4	4	195
HR-1007/8	3	282
HR-1009/10	3	214

3. The amniotic fluid or cervical DNA is considered to be of fetal origin if at least one VNTR/STR shows a different sized fragment when compared with the maternal DNA.

Comments

1. The four STR sequences should be analysed to provide sufficient data for exclusion analysis.

2. Always electrophorese the maternal and fetally-derived PCR amplified products in side-by-side lanes to assess VNTR/STR patterns for non-maternal origin.
3. Trinucleotide STR amplified sequences may show *Taq* polymerase-dependent amplification 'stutter' due to the variable addition of an adenine at the 3'-ends of the fragment.
4. The technique is used to confirm the origin of amniotic fluid or cervical smear DNA when a blood group compatible fetus is identified who is at risk for haemolytic disease of the newborn or neonatal alloimmune thrombocytopenia (see Denomme et al (1995) for further discussion).

Source

Denomme et al (1995).

3.17   Multiplex PCR Analysis

Principle

Simultaneous amplification of sequences using a single PCR reaction. This technique relies on sequence-specific amplification primers to detect two or more target sequences. The absence of an amplification product indicates that the sequence differs at or near the end of either sequence-specific primer and suggests the presence of a variant allele.

Specimens

Extracted DNA (see Protocol 3.6).

Equipment and Materials

Microcentrifuge
Thermal cycler
Micropipettors (general use)
Micropipettors (PCR use only)
Filter barrier micropipette tips

1.5 mL and 0.2 mL microcentrifuge tubes
10 × PCR buffer
25mM $MgCl_2$
100 mM dNTPs (Promega, Cat #U1240)
Nuclease-free ddH_2O (Promega Corp, Cat #P119C)
Allele-specific primers (100 ng/µL)
Control sense/antisense primers (100 ng/µL)
HotStarTaq DNA Polymerase (Qiagen Inc., Cat #203205)

Reagent Preparation

100 mM dNTPs are diluted to 10 mM using nuclease-free ddH_2O.

Procedure

1. Select the sequence-specific primers that are appropriate for the polymorphism under analysis (see ABO or Rh Facts Sheets in Part 4).
2. Prepare a PCR cocktail (see below) for the total number of samples to be analysed. Make up a volume that is 10% greater than the total volume required that includes the samples, controls and water blank.
3. Add DNA, vortex and flash spin tubes.
4. Amplify according to the multiplex PCR profile below.
5. Combine 10 µL of the PCR product with 2 µL of 6 × loading buffer (see Protocol 3.12) and electropherese using an appropriate gel system (see Part 2 for type and concentration of gel).
6. Photograph the gel. If unexpected bands are found, refer to the troubleshooting checklist in Part 2 (Table 2.4).

Multiplex PCR cocktail

Reagent	One sample	_____ samples
10 × buffer with 15 mM $MgCl_2$	5.0 µL	_____ µL
10 mM dNTPs	1.0 µL	_____ µL
Each of the sense primers	1.0 µL	_____ µL
Each of the antisense primers	1.0 µL	_____ µL
HotStarTaq polymerase (5.0 U/µL)	0.3 µL	_____ µL
Sterile ddH_2O	26.7 µL	_____ µL
Total	45.0 µL	_____ µL

Note: water volume defined for 12 primers. Adjust water accordingly when changing number of primers.

Combine 45 µL of either cocktail with 5 µL of DNA (1µg total) in a 0.2 mL microcentrifuge tube.

PCR profile

95°C for 15 minutes for HotStarTaq polymerase.
95°C for 60 seconds, 50 to 62°C for 60 seconds (see Procedural Note 3 below), 72°C for 45 seconds for 32 cycles.
72°C for 5 minutes, and 4°C hold.

Procedural Notes

1. The precautions listed in Protocol 3.8 will help avoid contamination.
2. The addition of *Taq* polymerase directly to the reaction mix along with DNA is possible for certain types of *Taq* polymerases. For example, the enzymatic activity of HotStarTaq polymerase is inhibited at room temperature but is activated upon heating above the annealing temperature. Therefore, mis-priming is avoided.
3. The PCR cocktail and profile are recommendations. The PCR conditions may need to be optimized (see Protocol 3.14).

Quality Control

1. The water blank (no DNA) must not have any bands.
2. Control DNA for each polymorphism must be analysed each time.

Results

1. All tests and control samples must have at least one of the bands amplified or an internal control band to validate the assay.
2. The test sample should have the corresponding band if the allele is present.

3.18   Southern Analysis

Principle

Genomic DNA is subjected to restriction enzyme digestion and size-separated by electrophoresis in an agarose gel. The DNA is transferred to a nylon membrane and analysed using digoxigenin (DIG)-11-dUTP labelled probes. Probes may be reverse transcribed cDNA, single exons, or larger regions of the gene. The hybridization of the DIG-labelled probe to the DNA is detected with anti-DIG/alkaline phosphatase and a chemiluminescent substrate. Chemiluminescence is detected by exposure of the membrane to X-ray film.

Specimens

Extracted DNA (Protocol 3.6).

Equipment and Materials

Microcentrifuge
Thermal cycler
Hybridization oven and bottles
DNA transfer apparatus, membrane and blotting paper (Scheleicher & Scheull TurboBlotter and Nytran membrane – 15 cm × 20 cm, Cat #416318)
UV cross-linker
X-ray cassette with intensifying screens (Eastman Kodak, Cat #KP 69065-E)
X-OMAT AR film (Eastman Kodak, Cat #165-1454)
Acetate page protectors
Micropipettors
PCR amplifed fragments:
  exon fragment
  exon–intron fragments
100 ng/µL primers for probe synthesis (for Rh analysis, see Part 4, Rh Facts Sheets)
DIG-11-dUTP Probe Synthesis Kit (Boehringer Mannheim, Cat #1636090)

DIG Luminescent and Detection Kit (Boehringer Mannheim, Cat #1363514)

DIG Easy Hybridization Solution (Boehringer Mannheim, Cat #1603558)

DIG Wash and Block Buffer Set (Boehringer Mannheim, Cat #1585762)

Sodium dodecyl sulphate (SDS; Sigma, Cat #L-4509)

Reagent Preparation

250 mM HCl

Denaturation solution (0.5 N NaOH/1.5 M NaCl)

Neutralization solution (0.5 M Tris–HCl, pH 7.5/3 M NaCl)

20 × SSC (3 M NaCl/300 mM sodium citrate, pH 7.0)

5 × SSC (1:4 (v:v) dilution of 20 × SSC in ddH$_2$O)

2 × Wash solution (2 × SSC containing 0.1% SDS)

0.5 × Wash solution (0.5 × SSC containing 0.1% SDS)

Procedure

A. Probe synthesis

1. Follow the manufacturer's instructions to prepare a PCR cocktail for DIG probe labelling. Use 5 µL of a 1:200 dilution of the gel-purified fragment as the DNA template.
2. Amplify the probe according to the PCR profile used to obtain the initial PCR fragment with the number of cycles reduced to 30. Amplify the labelling control according to the manufacturer's instructions.
3. To verify probe amplification efficiency, combine 5 µL of the PCR product with 1 µL of 6 × loading buffer and electrophorese using a 3% agarose gel (Protocol 3.11).
4. Photograph the gel under UV light. The expected PCR products should be clearly visible.

B. Southern blotting

1. Digest 5 µg of genomic DNA with the restriction enzyme of interest, (e.g. *Hae* III, *Msp* I, *Sph* I) according to the instructions provided with the enzyme.

2. Prepare a 1% agarose gel (15 cm × 25 cm) in TBE (Protocol 3.11). Dilute the digested DNA in 6 × DNA loading buffer (Protocol 3.11) and electrophorese until the bromophenol blue dye has migrated to within 5 cm of the bottom of the gel.
3. Place a fluorescent ruler next to the DNA ladder lane and photograph the gel under UV light.
4. Submerge the gel in denaturation solution twice for 15 minutes each. Rinse the gel with water.
5. Submerge the gel in neutralization solution twice for 15 minutes.
6. Prepare the nylon membrane for blotting according to the manufacturer's instructions.
7. Assemble the transblotting apparatus. Mark the position of the agarose lanes on the edge of the membrane using a pencil.
8. Transblot the DNA for a minimum of 6 hours in 20 × SSC.
9. Cross-link the DNA to the membrane using 2.4×10^5 millijoules of UV light. Rinse the membrane briefly in water and allow to air dry.

C.   Membrane Hybridization

1. Place the membrane in a hybridization bottle containing 10 mL of DIG Easy Hybridization Solution and cap. Pre-hybridize at 37–42°C for 2 hours with rotation in a hybridization oven.
2. Heat the DIG-labelled probe at 100°C for 10 minutes to denature the DNA. Chill on crushed ice.
3. Prepare at least 3.5 mL of DIG Easy Hybridization Solution plus the DIG-labelled probe for each 100 cm² membrane.
4. Discard the pre-hybridization solution. Add the DIG Easy Hybridization Solution containing the DIG-labelled probe (5–25 ng/mL). Allow the probe to hybridize overnight at 37–42°C.
5. Submerge the membrane in 2 × washing solution twice for 5 minutes each.
6. Submerge the membrane in 0.5 × washing solution twice for 5 minutes each.

D.   Chemiluminescent detection of DIG-label

1. Equilibrate the membrane in DIG Wash for 1 minute.
2. Allow the chemiluminescent substrate to come to room temperature.
3. Submerge the membrane in DIG Block Buffer for 30–60 minutes.

4. Dilute the anti-DIG-alkaline phosphatase 1:10,000 (v:v) using the Block Buffer.
5. Submerge the membrane in the diluted antibody solution for 30 minutes.
6. Discard the antibody solution. Submerge the membrane in DIG Wash twice for 15 minutes each.
7. Discard the DIG Wash and equilibrate the membrane in Detection Buffer for 2 minutes.
8. Dilute the substrate reagent provided in the kit 1:100 (v:v) using the Detection Buffer.
9. Place the membrane between two sheets of acetate plastic page protectors. Gently lift the top sheet and add approximately 0.5 mL of substrate per 100 cm² of membrane. Lower the top sheet of plastic. With a damp tissue, wipe the top sheet to remove any bubbles present under the sheet and to create a liquid seal around the membrane. Incubate for 5 minutes.
10. Transfer the semi-dry membrane to a plastic bag and seal the bag. Allow 7–8 hours for the chemiluminescence to reach steady state.
11. Incubate the membrane at 37°C for 15 minutes prior to exposure.
12. Expose the membrane to X-ray film. Multiple exposures from 20 minutes to 2 days are possible from a single blot.

Procedural Notes

A. Probe synthesis

1. See Protocol 3.8 for precautions to help avoid PCR contamination.
2. It is recommended to use a control template that can be DIG labelled (primers and template provided with the kit).
3. The semi-quantitative method (Protocol 3.7A) can be used to estimate the amount of DIG-labelled probe in place of the estimated yield using the spot test.
4. DIG-labelled DNA probes can be obtained also by random primary or nick translation.

B. Southern blotting

1. The photograph of the fluorescent ruler and agarose gel is used to determine the physical position of the DNA base pair ladder relative to the DNA fragments seen on the exposed X-ray film.

2. Do not allow the membrane to dry out until the DNA has been cross-linked.

C. Membrane hybridization

1. All incubations are at room temperature with shaking unless otherwise stated.
2. The DIG-labelled probe in the DIG Easy Hybridization Solution can be reused if desired. Store at −20°C for up to 1 year.

Quality Control

1. PCR contamination is detected by testing a water blank in place of the DNA template.
2. The chemiluminescent kits provide reagents to control for the DIG PCR labelling step.
3. Follow the manufacturer's instructions to estimate the yield of the DIG-labelled probe using a spot test and the DIG-labelled control.

Results

DNA will vary in their Southern restriction maps depending on the zygosity and inherited polymorphisms for the gene of interest.

Comments

1. Common genotypes can be used to obtain restriction maps.
2. Mapping a gene locus using Southern analysis and many restriction enzymes can differentiate certain haplotypes and is a useful tool for the study of variants or antigen-negative phenotypes with a discrepant DNA-based genotype.

Source

Ausubel et al (1997); Southern (1975).

3.19   References

Ausubel, F.M., Brent, R., Kingston, R.E., et al (1997) *Current Protocols in Molecular Biology* (3 volumes). John Wiley & Sons, New York.

Denomme, G. and Fernandes, B.J. (1999) The assignment of *RHD* gene zygosity by densitometric quantification. Abstract. *Transfusion* 39 (Suppl 1) 51S.

Denomme, G.A., Waye, J.S., Burrows, R.F., et al (1995) The prenatal identification of fetal compatibility in neonatal alloimmune thrombocytopenia using amniotic fluid and variable number of tandem repeat (VNTR) analysis. *Br J Haematol* 91, 742–746.

Ellis, W.D., Mulvaney, B.D. and Saathoff, D.J. (1975) Leukocyte isolation by sedimentation: The effect of rouleau-promoting agents on leukocyte differential count. *Prep Biochem* 5, 179–187.

Epstein, C.J. (1982) The use of growth factors in stimulate the proliferation of amniotic fluid cells. *Methods Cell Biol* 26, 269–276.

Hecht, F., Peakman, D.C., Kaiser-McCaw, B., et al (1981) Amniocyte clones for prenatal cytogenetics. *Am J Med Genet* 10, 51–54.

Orita, M., Iwahana, H., Kanazawa, H., et al (1989a) Detection of polymorphisms of human DNA by gel electrophoresis as single-strand conformation polymorphisms. *Proc Natl Acad Sci USA* 86, 2766–2770.

Orita, M., Suzuki, Y., Sekiya, T., et al (1989b) Rapid and sensitive detection of point mutations and DNA polymorphisms using the polymerase chain reaction. *Genomics* 5, 874–879.

Sambrook, J., Fritsch, E.F. and Maniatis, T. (1989) *Molecular Cloning: A Laboratory Manual* (3 volumes), 2nd edition. Cold Spring Harbor Laboratory Press, Cold Spring Harbor, NY.

Southern, E.M. (1975) Detection of specific sequences among DNA fragments separated by gel electrophoresis. *J Mol Biol* 98, 503–517.

Tellez, A. and Rubinstein, P. (1970) Rapid method for separation of blood cells. *Transfusion* 10, 223–225.

Part 4   Red Cell Blood Groups

4.1   Terminology

The terminology used for antigens is that recommended by the Working Party on Terminology for Red Cell Surface Antigens of the International Society of Blood Transfusion (ISBT) (Daniels et al 1995, 1996, 1999). This Working Party has categorized over 250 different blood group antigens into Systems, Collections, Low Incidence Antigens, and High Incidence Antigens. The terminology is inconsistent. A single letter (e.g., A, D, K), a symbol and a superscript (e.g., Fy^a, Fy^b, Jk^a), and a numerical notation (e.g., Fy 3, Lu 4, K 11) are used even within the same system (e.g., Fy^a, Fy3). A blood group 'system' consists of one or more antigens controlled by a single gene locus, or by two or more closely linked homologous genes with little or no observable recombination between them.

4.2   Clinical Applications

Blood group antigens have relevance in transfusion medicine because people whose cells lack an antigen can be alloimmunized if they are exposed to the corresponding antigen. This can occur in the setting of transfusion or fetal–maternal incompatibility, and alloantibodies can cause destruction of antigen-positive RBCs *in vivo*. Classical haem-agglutination is specific, sensitive, quick to perform, and inexpensive but has limitations. Molecular analysis can be used to overcome these serological limitations.

Table 4.1

Red cell blood group systems

Blood Group System			Gene Name		Molecular
Classical	ISBT No.	ISBT Symbol	ISBT	HGM	protocols provided
ABO	001	ABO	*ABO*	*ABO*	
MNS	002	MNS	*MNS*	*GYPA*	*MNS 1/MNS 2*
				GYPB	*MNS 3/MNS 4*
P	003	P1	*P1*	*P1*	
Rh	004	RH	*RH*	*RHD*	*RH 1*
				RHCE	*RH 3/RH 5*
Lutheran	005	LU	*LU*	*LU*	*LU 1/LU 2*
Kell	006	KEL	*KEL*	*KEL*	*KEL 1/KEL 2*
					KEL 6/KEL 7
Lewis*	007	LE	*LE*	*FUT3*	
Duffy	008	FY	*FY*	*DARC*	*FY 1/FY 2 GATA1;*
					nt265; nt298
Kidd	009	JK	*JK*	*SLC14A1*	*JK 1/JK 2*
Diego	010	DI	*DI*	*SCL4A1*	
Yt	011	YT	*YT*	*ACHE*	
Xg	012	XG	*XG*	*XG*	
Scianna	013	SC	*SC*	*SC*	
Dombrock	014	DO	*DO*	*DO*	
Colton	015	CO	*CO*	*AQP1*	
Landsteiner–Wiener	016	LW	*LW*	*LW*	
Chido/Rodgers*	017	CH/RG	*CH/RG*	*C4A, C4B*	
Hh	018	H	*H*	*FUT1*	
Kx	019	XK	*XK*	*XK*	
Gerbich	020	GE	*GE*	*GYPC*	
Cromer	021	CROM	*CROM*	*DAF*	
Knops	022	KN	*KN*	*CR1*	
Indian	023	IN	*IN*	*CD44*	
Ok	024	OK	*OK*	*CD147*	
Raph	025	RAPH	*MER2*	*MER2*	

* Lewis and Chido/Rogers are not included in section 4.5

4.2.1   Identification of a Fetus at Risk for Anaemia of the Neonate (Haemolytic Disease of the Newborn (HDN))

Haemagglutination studies, including antibody titres, give only an indirect measure of potential complications for HDN. Molecular genotyping overcomes the serological limitation to predict at-risk pregnancies for HDN. Sufficient DNA can be prepared from amniocytes obtained from routine amniocentesis as early as 13 weeks gestation. Fetal DNA can also be obtained from chorionic villi, cervical smear, or maternal plasma, thereby avoiding invasive procedures (Bennett et al 1993, 1995; Kingdom et al 1995; Lo et al 1998).

We recommend that the decision to perform molecular analyses for clinical use be based on the following:

- Mother has an alloantibody of potential clinical significance for HDN.
- Father is heterozygous for the allele or is unknown.
- DNA genotyping and RBC phenotyping is performed on the parents. This will limit the gene pool and can identify variants that could influence the interpretation of results from the fetus.
- The ethnicity of the parents is obtained. This can help focus the test approach because some variants are restricted to certain populations. For example, in black Africans, *FY 2* in tandem with a mutation in GATA1 causes non-translation of the Fy^b protein in the erythroid lineage (Chaudhuri et al 1995; Tournamille et al 1995; Iwamoto et al 1996).
- The results are validated by antigen typing the baby's RBCs.

4.2.2   Determination of the Presence or Absence of Blood Group Alleles in Recently Transfused Patients

DNA can be obtained from WBCs in the patient's peripheral blood, or from cells obtained by non-invasive procedures such as buccal smear and urine sediment (Rios et al 1999). We have observed that the WBCs remaining in transfused blood products (Wenk and Chiafari 1997; Lee et al 1995; Adams et al 1992) do not affect the assays given here (Reid et al 1999a). Cross-contamination between blood samples through probing on clinical automated machines is unlikely to be detected unless more sensitive methods are used (Reed et al 1997; Cheung et al 1997; Lee et al 1996).

4.2.3   Determination of Blood Group Alleles when Patients' RBCs are Coated with Immunoglobulin

Molecular genotyping is helpful if a method used to remove the antibody from RBCs denatures the target antigen (e.g., the acid elution method destroys antigens in the Kell blood group system (Julleis et al 1992)) or when a chemical treatment does not remove sufficient IgG to give confident results. This is especially true when only weakly reactive antibodies are available.

4.2.4   Resolution of A, B and D Typing Discrepancies

RBCs that express weak subgroups of A or B can result in blood group typing discrepancies. This occurs when an altered gene encodes for a transferase that adds low levels of an A (or B) immunodominant sugar to the precursor H antigen. Since the molecular bases of many of the weak subgroups of A and B are associated with altered transferase genes, PCR-RFLP can be used to define the transferase gene and, thus, the ABO group.

Anti-D reagents contain monoclonal antibodies that detect small and specific parts of the D antigen. Some donors and patients who in the past have typed as D-negative are now typed as D-positive. This occurs when monoclonal antibodies detect epitopes that were not detected by polyclonal reagents. Genotyping can identify the molecular basis for these apparent discrepancies.

4.2.5   Screening for Antigen-negative Blood Donors

Genotyping is useful when antisera are weakly reactive or are not available in sufficient quantity to screen large numbers of donor samples. However, mass screening of blood donors will be possible when DNA microchip or similar technology is available for genotyping in a rapid, throughput system (Schafer and Hawkins 1998).

4.2.6   *RHD* Zygosity

RHD zygosity assessment by phenotypic analysis provides only a 'most probable' genotype for the D-positive individual with other possible genotypes including both homozygous and hemizygous haplotypes. Since the D-negative genotype is generally characterized by the absence of *RHD*, quantitative PCR amplification of the *RHD* and *RHCE* (Protocol 3.15) can be used to determine whether the genome of a D-positive individual has one or two copies of the *RHD*. Table 4.2 provides exon 7 and exon 10 primers used to assess genomic DNA for *RHD* zygosity. The assay uses densitometric scanning of PCR amplified fragments separated by gel electrophoresis. However, it is important to note that other detection systems could be developed (e.g. fluorescent PCR amplification).

Table 4.2

Rh primers for *RHD* zygosity assessment (Bennett et al 1994)

Primer name	Region	Sequence (5' to 3')	Gene homology	
			RHD	*RHCE*
Rh7s	Exon 7	TGTGTTGTAACCGAGT	Yes	Yes
Rh7a	Exon 7	ACATGCCATTGCCG	Yes	Yes
Rh10s	Exon 10	TAAGCAAAAGCATCCA	Yes	Yes
RhDa	3' UTR	ATGGTGAGATTCTCCT	Yes	No

UTR, untranslated region.

4.3   Situations where Genotype and Phenotype do not Agree

There are several situations where genotype and phenotype results are apparently discrepant (Reid et al 1999b). Some are listed below.

- Chronic/massive transfusions where the phenotype of the patient's RBC sample may be that of transfused RBCs.
- Changes in the gene that affect either its transcription (e.g., point mutation in GATA1 of *FY*), splicing (point mutation of splice motif, e.g., *JK*, *GYPA*), introduction of a stop codon (e.g., *RHD*, *FY*, *CO*, *GYPC*, *DAF*), or stability of the membrane protein (e.g., point mutation in *CO*).
- Crossovers and other gene rearrangements (e.g., *RHCE/RHD*, *GYPA/GYPB*).
- Changes in a modifying gene (e.g., *IN(LU)*, *IN(JK)*)
- Changes in a different gene whose product is required for proper expression of the protein carrying the blood group antigen in the RBC membrane (e.g., *4.1* and Gerbich antigens; *RHAG* and Rh antigens; *GYPA* and Wr[b] antigen; *XK* and Kell antigens).
- Medical procedures, e.g., transplantation with homologous stem cells, artificial insemination or surrogate motherhood.

Occasionally, blood group genotyping may not correlate with the phenotype in instances when there is no evidence of recent transfusion or transplantation. Both allele-specific multiplex PCR (Protocol 3.17) and Southern analysis (Protocol 3.18) are useful techniques for studying gene structure anomalies. Allele-specific multiplex PCR can provide a quick and easy 'snapshot' of the presence or absence of mutated or variant alleles (see Table 4.3 for ABO, and Table 4.4 for Rh). On the other hand, using several restriction enzymes and various 5' or 3' gene-specific probes in Southern analysis, it may be possible to identify genetic alterations to a gene (e.g., inversions, deletions, insertions). Information for the *RH* loci is available, although virtually any gene locus could be analysed once labelled probes have been obtained. Table 4.5 provides a list of *RH* primers used to obtain probes for hybridization in Southern analysis.

Table 4.3

Allele-specific primers for ABO multiplex PCR amplification (Gassner et al 1996)

Allele	Sense primer (5' to 3')	Antisense primer (5' to 3')
O¹	TTAAGTGGAAGGATGTCCTCGTCGTA	ATATATATGGCAAACACAGTTAACCCAATG
Non-O¹	TAAGTGGAAGGATGTCCTCGTCGTG	
O²		TCGACCCCCCGAAGAAGCT
non-O²	AGTGGACGTGGACATGGAGTTCC	CGACCCCCCGAAGAAGCC
B		ATCGACCCCCCGAAGAGCG
non-B		CCGACCCCCCGAAGAGCC
A²	GAGGCGGTCCGGAAGCG	GGGTGTGATTTGAGGTGGGGAC
non-A²	GAGGCGGTCCGGAACACG	
Control	TGCCTTCCCAACCATTCCCTTA	CCACTCACGGATTTCTGTTGTGTTTC

Table 4.4

Allele-specific primers for *RHD* multiplex PCR amplification (Maaskant-van Wijk et al 1998). Correction (Primer Exon 6) Maaskant-van Wijk et al 1999

Region	Sense primer (5' to 3')	Antisense primer (5' to 3')
Exon 3	TCGGTGCTGATCTCAGTGGA	ACTGATGACCATCCTCATGT
Exon 4	CACATGAACATGATGCACA	CAAACTGGGTATCGTTGCTG
Exon 5	GTGGATGTTCTGGCCAAGTT	CACCTTGCTGATCTTACC
Exon 6	GTGGCTGGGCTGATCTACG	TGTCTAGTTTCTTACCGGCAAGT
Exon 7	AGCTCCATCATGGGCTACAA	ATTGCCGGCTCCGACGGTATC
Exon 9	AACAGGTTTGCTCCTAAATATT	AAACTTGGTCATCAAAATATTTAACCT

Table 4.5

PCR amplification primers for hybridization probes used in Southern analysis
(Hyland et al 1994; Denomme et al 1999)

Region	Sense primer (5' to 3')	Antisense primer (5' to 3')
Rh UTR – exon 1	GATCTGTTCCTTGCTTTTCTTACAAGG	CTCTAAGGAAGCGTCATAGTGGGTAAA
RHD exons 4–5	ACGATACCCAGTTTGTCT	TGACCCTGAGATGGCTGT
RHD exon 10 – UTR	TAAGCAAAAGCATCCA	ATGGTGAGATTCTCCT

4.4   References

Adams, P.T., Davenport, R.D., Reardon, D.A., et al (1992) Detection of circulating donor white blood cells in patients receiving multiple transfusions. *Blood* 80, 551–555.

Bennett, P.R., Le Van Kim, C., Colin, Y., et al (1993) Prenatal determination of fetal RhD type by DNA amplification. *N Engl J Med* 329, 607–610.

Bennett, P.R., Overton, T.G., Lighten, A.D., et al (1995) Rhesus D typing. *Lancet* 345, 661–662.

Chaudhuri, A., Polyakova, J., Zbrzezna, V., et al (1995) The coding sequence of Duffy blood group gene in humans and simians: Restriction fragment length polymorphism, antibody and malarial parasite specificities, and expression in non-erythroid tissues in Duffy-negative individuals. *Blood* 85, 615–621.

Cheung, M.-C., Goldberg, J.D. and Kan, W.K. (1997) Prenatal diagnosis of sickle cell anemia and thalassaemia by analysis of fetal cells in maternal blood. *Nature Genet* 14, 264–268.

Daniels, G.L., Anstee, D.J., Cartron, J.-P., et al (1995) Blood group terminology 1995. ISBT working party on terminology for red cell surface antigens. *Vox Sang* 69, 265–279.

Daniels, G.L., Anstee, D.J., Cartron, J.P., et al (1996) Terminology for red cell surface antigens – Makuhari report. *Vox Sang* 71, 246–248.

Daniels, G.L., Anstee, D.J., Cartron, J.P., et al (1999) Terminology for red cell surface antigens – Oslo report. *Vox Sang* 77, 52–57.

Denomme, G.A., Akoury, H., Sermer, M., et al (1999) RhD status of a fetus at risk for haemolytic disease with a discrepant maternal DNA-based RhD genotype. *Prenat Diagn* 19, 424–427.

Gassner, C., Schmarda, A., Nussbaumer, W., et al (1996) ABO glycosyltransferase genotyping by polymerase chain reaction using sequence-specific primers. *Blood* 88, 1852–1856.

Hyland, C.A., Wolter, L.C., Liew, Y.W., et al (1994) A Southern analysis of Rh blood group genes: association between restriction fragment length polymorphism patterns and Rh serotypes. *Blood* 83, 566–572.

Iwamoto, S., Li, J., Sugimoto, N., et al (1996) Characterization of the Duffy gene promoter: Evidence for tissue-specific abolishment of expression in Fy(a-b-) of black individuals. *Biochem Biophys Res Commun* 222, 852–859.

Julleis, J., Sapp, C. and Kakaiya, R. (1992). Glycine-EDTA as a substitute for AET in the inactivation of Kell system antigens on red blood cells. Abstract. *Transfusion* 32(Suppl), 14S.

Kingdom, J., Sherlock, J., Rodeck, C., et al (1995) Detection of trophoblast cells in transcervical samples collected by lavage or cytobrush. *Obstet Gynecol* 86, 283–288.

Lee, T.-H., Donegan, E., Slichter, S., et al (1995) Transient increase in circulating donor leukocytes after allogeneic transfusions in immunocompetent recipients compatible with donor cell proliferation. *Blood* 85, 1207–1214.

Lee, T.-H., Paglieroni, T., Ohro, H., et al (1996). Longterm multi-lineage chimerism of donor leukocytes in transfused trauma patients. Abstract. *Blood* 88 (Suppl 1), 265a.

Lo, Y.M.D., Hjelm, N.M., Fidler, C., et al (1998) Prenatal diagnosis of fetal RhD status by molecular analysis of maternal plasma. *N Engl J Med* 339, 1734–1738.

Maaskant-van Wijk, P.A., Faas, B.H., De Ruijter, J.A., et al (1998) Genotyping of *RHD* by multiplex polymerase chain reaction analysis of six *RHD*-specific exons. *Transfusion* 38, 1015–1021. Maaskant-van Wijk et al (1999) Correction. *Transfusion* 39, 546.

Reed, W., Lee, T.-H., Busch, M.P., et al (1997). Sample suitability for the detection of minor leukocyte populations by polymerase chain reaction (PCR). Abstract. *Transfusion* 37 (Suppl), 107S.

Reid, M.E., Rios, M., Powell, V.I., et al (1999a) DNA from blood samples can be used to genotype patients who have been recently transfused. *Transfusion*, in press.

Reid, M.E., Rios, M. and Yazdanbakhsh, K. (1999b) Applications of molecular biology techniques to transfusion medicine. *Semin Hematol*, in press.

Rios, M., Cash, K., Strupp, A., et al (1999) DNA from urine sediment or buccal cells can be used for blood group molecular genotyping. *Immunohematology* 15, 61–65.

Schafer, A.J. and Hawkins, J.R. (1998) DNA variation and the future of human genetics. *Nat Biotechnol* 16, 33–39.

Tournamille, C., Colin, Y., Cartron, J.P., et al (1995) Disruption of a GATA motif in the *Duffy* gene promoter abolishes erythroid gene expression in Duffy-negative individuals. *Nature Genet* 10, 224–228.

Wenk, R.E. and Chiafari, F.A. (1997) DNA typing of recipient blood after massive transfusion. *Transfusion* 37, 1108–1110.

4.5   Facts Sheets, Gene Maps and Molecular Protocols

ABO Blood Group System
Facts Sheet

ISBT Gene Name:	*ABO*
Organization:	7 exons distributed over >18 kbp (cDNA ~1,580 bp)
Chromosome:	9q34.1–q34.2
Gene product:	A: α1-3-*N*-acetyl-D-galactosaminyltransferase (353 amino acids)
	B: α1-3-D-galactosaminyltransferase (353 amino acids)
GenBank Accession:	AF134412-44; D82825-28

A

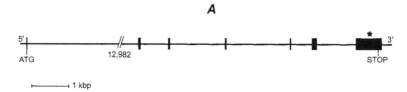

Figure 4.1 *ABO* gene. For correlation of restriction enzymes and selected alleles, see Table 4.6. *Most variants are encoded by missense mutations in exon 7; see Tables 4.7, 4.8, 4.9.

Alleles:	*ABO 1/ABO 2* and numerous variants
Antigens:	A, B and numerous subtypes

Prevalence of gene products (% based on phenotyping):

Phenotype	Caucasians	Blacks	Asians
A	40	27	28
B	11	20	27
AB	4	4	4
O	45	49	41

Molecular Details

Table 4.6
Correlation of restriction enzymes and selected alleles

Allele detected	Exon	Nucleotide mutation	Restriction endonuclease
O^{1var}	4	188G>A; 189C>T	*BstU* I
O^1, O^{1var}	6	261G deleted	*Kpn* I
A^2, O^3, cis-AB	7	467C>T	*Hpa* II
B, O^2	7	526C>T	*Nar* I/*BssH* II
A^x, O^{1var}	7	646T>A	*Mbo* I
B	7	703G>A	*Alu* I/*Hap* II
A^x, O^{1var}	7	771C>T	*Dde* I
A^3	7	871G>A	*Sal* I
B, O^2	7	1096G>A	*Hpa* II
A^{el}, O^3	7	798G inserted	AS-PCR
A^2, O^3	7	1059C deleted	AS-PCR

For corresponding amino acids, see Tables 4.7 to 4.9.

Table 4.7

A alleles

			Exon 6		Exon 7																				
Phenotype	Allele	Other Name	nt: 261	297	467	526	564	641	646	657	669	681	703	721	771	796	802	803	798-804	829	871	930	1009	1054	1059-1061
			aa: –	–	156	176	–	214	216	–	223	–	235	–	–	266	267	268	–	277	291	–	337	352	–
A₁	ABO A101	A¹	G	A	C Pro	C Arg	C	T Met	T Phe	C	G Glu	G	G Gly	C	C	C Leu	G Gly	G Gly	GGGGGGG	G Val	G Asp	G	G	C Arg	CCC
A₁	ABO A102	A¹	–	–	T Leu	–	–	–	–	–	–	–	–	–	–	–	–	–	–	–	–	–	A Arg	–	–
A₁	ABO A103	–	–	–	T Leu	–	T	–	–	–	–	–	–	–	–	–	–	–	–	–	–	–	–	–	–
A₁	ABO A104	–	–	G	–	–	–	–	–	–	–	–	–	–	–	–	–	–	–	–	–	–	–	–	–
A₂	ABO A105	A²	–	–	T Leu	–	–	–	–	–	–	–	–	–	–	–	–	–	–	–	–	–	–	→	CC†
A₂	ABO A106	–	–	–	–	–	–	–	–	–	–	–	–	–	–	–	–	–	–	–	–	–	–	–	–
A₂	ABO A107	–	–	–	–	–	–	–	–	–	–	–	–	–	–	–	–	–	–	–	–	–	T Trp	G Gly	–
A₂	ABO A111	–	–	–	T Leu	–	–	–	–	–	–	–	–	–	–	–	–	–	–	–	–	–	–	–	–
A₃	–	A³	–	–	–	–	–	–	–	–	–	–	–	–	–	–	–	–	–	–	–	–	G Gly	–	–
Aₓ	ABO A108	Aˣ	–	–	–	–	–	–	A Ile	–	–	–	–	–	–	–	–	–	–	–	A Asn	–	–	–	–
A_el	ABO A109	–	–	–	–	–	–	–	–	–	–	A	–	–	–	–	–	–	–	–	–	–	–	–	–
A_el	ABO A110	–	–	–	T Leu	–	–	–	A Ile	T	–	–	–	–	T	–	–	–	–	–	–	–	–	–	–
A₂	ABO R101	–	–	–	–	G Gly	–	–	–	–	–	–	A Ser	–	–	–	–	–	–	A Met	–	–	–	–	–
Cis-AB	ABO C101	–	–	–	T Leu	–	–	–	–	–	–	–	–	–	–	–	–	C Ala	–	–	–	–	–	–	–
A	–	A^Sromcek	–	–	T Leu	G Gly	–	–	–	–	–	–	–	–	–	–	–	–	–	–	–	–	–	–	–

† Deletion of nt 1059(C) and then an extra 21 aa.

Table 4.8

B alleles

Phenotype	Allele	Other Name	Exon 6		Exon 7																					
			nt: 261	297	467	526	564	641	646	657	669	681	703	721	771	796	802	803	798–804	829	871	930	1054	1059–1061	1096
			aa: –	–	156	176	–	214	216	–	223	–	235	–	–	266	267	268	–	277	291	–	352	–	–
A₁	ABO A101	A¹	G	A	C (Pro)	C (Arg)	C	–	T (Phe)	C	G (Glu)	G	G (Gly)	C	C	C (Leu)	G (Gly)	G (Gly)	GGGGGGG	G (Val)	G (Asp)	G	C (Arg)	CCC	G
Cis-AB	ABO C101	–	–	–	T (Leu)	–	–	–	–	–	–	–	–	–	–	–	–	–	–	–	–	–	–	–	–
B	ABO B101	–	–	G	–	G (Gly)	–	–	–	T	–	–	A (Ser)	–	–	A (Met)	–	C (Ala)	–	–	–	A	–	–	–
B	ABO B102	–	–	G	–	G (Gly)	–	–	–	T	–	–	A (Ser)	–	–	A (Met)	–	C (Ala)	–	–	–	A	–	–	–
B	ABO B103	–	–	G	–	G (Gly)	–	–	–	T	–	–	A (Ser)	–	–	A (Met)	–	C (Ala)	–	–	–	A	–	–	–
B	ABO B107	–	–	G	–	G (Gly)	–	–	–	T	–	–	A (Ser)	–	–	A (Met)	–	C (Ala)	–	–	–	A	–	–	–
Bₓ	ABO B104	–	–	G	–	G (Gly)	–	–	–	T	–	–	A (Ser)	–	–	A (Met)	–	C (Ala)	–	–	A (Asn)	A	–	–	–
B_el	ABO B105	–	–	G	–	G (Gly)	–	G (Arg)	–	T	–	–	A (Ser)	–	–	A (Met)	–	C (Ala)	–	–	–	A	–	–	–
B_el	ABO B106	–	–	G	–	G (Gly)	–	–	–	T	T (Asp)	–	A (Ser)	–	–	A (Met)	–	C (Ala)	–	–	–	A	–	–	–
B₃		–	–	G	–	G (Gly)	–	–	–	T	–	–	A (Ser)	–	–	A (Met)	–	C (Ala)	–	–	–	–	T (Trp)	–	–
B₃		B³	–	G	–	G (Gly)	–	–	–	T	–	–	A (Ser)	–	–	A (Met)	–	C (Ala)	–	–	A	–	T (Trp)	–	–
B_(A)		–	–	–	–	G (Gly)	–	–	–	–	–	–	A (Ser)	–	–	A (Met)	–	C (Ala)	–	–	–	–	–	–	–
B_(A)		B^(A)	–	G	–	G (Gly)	–	–	–	–	–	–	A (Ser)	–	–	A (Met)	–	C (Ala)	–	–	A	–	–	–	A

Table 4.9
O alleles

			Exon 6			Exon 7																				
Phenotype	Allele	Other Name	nt: 261 aa: –	297 –	454 –	467 156	526 176	564 –	641 214	646 216	657 –	669 223	681 –	703 235	721 –	771 –	796 266	802 267	803 268	798–804 –	829 277	871 291	930 –	1054 352	1059–1061 –	1096 –
A₁	ABO A101	A¹	G	A	T	C Pro	C Arg	C	T Met	T Phe	C	G Glu	G	G Gly	C	C	C Leu	G Gly	G Gly	GGGGGGG	G Val	G Asp	G	C Arg	CCC	G
O	ABO O101	O¹	ΔG↑	–	STOP aa116	–	–	–	–	–	–	–	–	–	–	–	–	–	–	–	–	–	–	–	–	–
O	ABO O102	O¹	ΔG↑	–	STOP	–	–	C	–	–	–	–	–	–	–	–	–	–	–	–	–	–	–	–	–	–
O	ABO O103	O¹	ΔG↑	G	STOP	–	–	–	–	–	–	–	–	–	–	–	–	–	–	–	–	–	–	–	–	–
O	ABO O104	O¹	ΔG↑	–	C	–	–	–	–	–	–	–	–	–	–	–	–	–	–	–	–	–	–	–	–	–
O	ABO O201	O¹ variant	ΔG↑	G	STOP	–	–	–	–	A	–	–	A	–	–	T	–	–	–	–	A	–	–	–	–	–
O	ABO O202	O¹	ΔG↑	–	STOP	–	–	–	–	A	–	–	A	–	–	T	–	–	–	–	A	–	–	–	–	–
O	ABO O203	O¹	ΔG↑	G	STOP	–	–	–	–	A	–	–	A	–	T	T	–	–	–	–	A	–	–	–	–	–
O	ABO O2	O²	–	–	G	–	G Gly	–	–	–	–	–	–	–	–	–	–	A Arg	–	–	–	–	–	–	–	A

References

Gassner, C., Schmarda, A., Nussbaumer, W., et al (1996) ABO glycosyltransferase geno-typing by polymerase chain reaction using sequence-specific primers. *Blood* 88, 1852–1856.

Grunnet, N., Steffensen, R., Bennett, E.P., et al (1994) Evaluation of histo-blood group ABO genotyping in a Danish population: frequency of a novel O allele defined as O^2. *Vox Sang* 67, 210–215.

Ogasawara, K., Bannai, M., Saitou, N., et al (1996) Extensive polymorphism of ABO blood group gene: Three major lineages of the alleles for the common ABO pheno-types. *Hum Genet* 97, 777–783.

Ogasawara, K., Yabe, R., Uchikawa, M., et al (1996) Molecular genetic analysis of variant phenotypes of the ABO blood group system. *Blood* 88, 2732–2737.

Ogasawara, K., Yabe, R., Uchikawa, M., et al (1998) Different alleles cause an imbalance in A_2 and A_2B phenotypes of the ABO blood group. *Vox Sang* 74, 242–247.

Olsson, M.L. and Chester, M.A. (1995) A rapid and simple ABO genotype screening method using a novel B/O^2 versus A/O^2 discriminating nucleotide substitution at the ABO locus. *Vox Sang* 69, 242–247.

Olsson, M.L. and Chester, M.A. (1996) Evidence for a new type of O allele at the ABO locus, due to a combination of the A^2 nucleotide deletion and the A^{el} nucleotide insertion. *Vox Sang* 71, 113–117.

Olsson, M.L. and Chester, M.A. (1996) Frequent occurrence of a variant O^1 gene at the blood group ABO locus. *Vox Sang* 70, 26–30.

Stroncek, D.F., Konz, R., Clay, M.E., et al (1994) Determination of ABO glycosyl-transferase genotypes by use of polymerase chain reaction and restriction enzymes. *Transfusion* 35, 231–240.

Yamamoto, F. (1995) Molecular genetics of the ABO histo-blood group system. *Vox Sang* 69, 1–7.

Yamamoto, F., Clausen, H., White, T., et al (1990) Molecular genetic basis of the histo-blood group ABO system. *Nature* 345, 229–233.

MNS Blood Group System
Facts Sheet

ISBT Gene Name: *MNS (GYPA/GYPB)*

Organization: *GYPA*: 7 exons distributed over 60 kbp (cDNA 2,591 bp)
GYPB: 5 exons (plus 1 pseudoexon) distributed over 58 kbp (cDNA 612 bp)

Chromosome: 4q28.2–q31.1

Gene product: Glycophorin A (131 amino acids)
Glycophorin B (72 amino acids)

GenBank Accession: *GYPA*: X51798; M60707
GYPB: JO2982; M60708

MNS

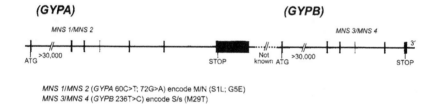

MNS 1/MNS 2 (*GYPA* 60C>T; 72G>A) encode M/N (S1L; G5E)
MNS 3/MNS 4 (*GYPB* 236T>C) encode S/s (M29T)

⊢────────┤ 1 kbp

Figure 4.2   *MNS* gene.

Alleles: *MNS* 1/*MNS* 2; *MNS* 3/*MNS* 4

Antigens: M/N; S/s and numerous variants

Prevalence of gene products (% based on phenotyping):

Phenotype	Caucasians	Blacks
M+N–S+s–	6	2
M+N–S+s+	14	7
M+N–S–s+	8	16
M+N+S+s–	4	2
M+N+S+s+	24	13

M+N+S–s+	22	33
M–N+S+s–	1	2
M–N+S+s+	6	5
M–N+S–s+	15	19
M+N–S–s–	0	0.4
M+N+S–s–	0	0.4
M–N+S–s–	0	0.7

Molecular protocols

PCR Condition: Cocktail in Protocol 3.10
Thermal Cycler: Profile 1 in Protocol 3.10

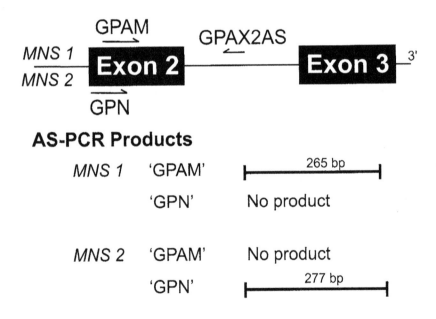

Internal control (531 bp with RHCE primers) must be present in all reactions.

Figure 4.3   *MNS 1* (M)/*MNS 2* (N), AS-PCR.

PCR Condition: Cocktail in Protocol 3.10
Thermal Cycler: Profile 1 in Protocol 3.10

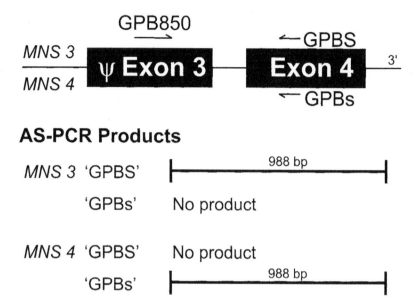

AS-PCR Products

MNS 3 'GPBS' |————————— 988 bp —————————|

'GPBs'   No product

MNS 4 'GPBS'   No product

'GPBs' |————————— 988 bp —————————|

Internal control (531 bp with RHCE primers)
must be present in all reactions.

Figure 4.4   *MNS 3* (S)/*MNS 4* (s), AS-PCR.

Primers

Name	Sequence	GenBank Accession
GPA M	5'-TATCAGCATCAAGTACCACTGG-3'	M60707
GP N	5'-AAATTGTGAGCATATCAGCATTA-3'	M60707
GPA X2AS	5'-TCAGAGGCAAGAATTCCTCCA-3'	M60707
GPB S	5'-AGTGAAACGATGGACAAGTTCTCCCA-3'	M60708
GPB s	5'-AGTGAAACGATGGACAAGTTCTCCCG-3'	M60708
GPB 850	5'-GTTTCCCCTCCAGAAAAGAAAAACGT-3'	M60708

Internal control primers

RHCES	5'-CCTATTTTGCGCTGTACTGTGG-3'	Y10604
RHCEAS	5'-GCTCACTGCCCGATTTTCTAT-3'	Y10604

References

Fukuda, M. (1993) Molecular genetics of the glycophorin A gene cluster. *Semin Hematol* 30, 138–151.

Siebert, P.D. and Fukuda, M. (1987) Molecular cloning of a human glycophorin B cDNA: Nucleotide sequence and genomic relationship to glycophorin A. *Proc Natl Acad Sci USA* 84, 6735–6739.

Rh Blood Group System
Facts Sheet

ISBT Gene Name:	*RH (RHCE; RHD)*
Organization:	*RHCE:* 10 exons distributed over 69 kbp (cDNA 1,455 bp)
	RHD: 10 exons distributed over 69 kbp (cDNA 1,354 bp)
Chromosome:	1p36.13–p34.3
Gene product:	RhCE protein (417 amino acids)
	RhD protein (417 amino acids)

GenBank Accession:

RH:	U66340; U66341; X54534; Z97364; L08429; M34015
RHD:	Intron 1 (part) Z97363; intron 3 (part) Z97031; intron 4 Y10605; introns 5 and 6 Z97334.
RHCE:	Intron 1 (part) Z97362; intron 3 (part) Z97030; intron 4 Y10604; introns 5 and 6 Z97333.
Alleles:	*RHce; RHCe; RHcE; RHCE*
	RHD
Antigens:	C/c, E/e and numerous variants
	D and numerous variants. The D antigen consists of many epitopes (epD) along the RhD protein.

Prevalence of gene products (% based on phenotyping):

Haplotype	Caucasians	Blacks	Asians
ce	37	26	3
DCe	42	17	70
DcE	14	11	21
Dce	4	44	3
Ce	2	2	2
cE	<1	<1	<1
DCE	<1	<1	1
CE	<1	<1	<1

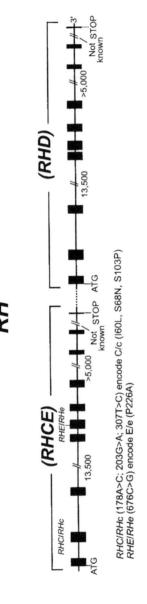

Figure 4.5 *RH* gene.

Molecular Protocols

PCR Condition: Cocktail A or C in Protocol 3.8
Thermal Cycler: Profile 1 in Protocol 3.8

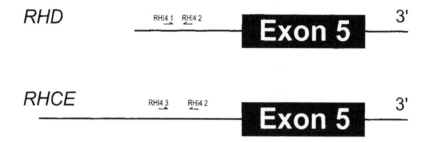

PCR Product

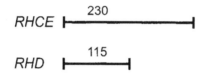

Figure 4.6   *RH* (Rh D/Rh nonD) in intron 4, AS-PCR.

PCR Condition: Cocktail A in Protocol 3.8
Thermal Cycler: Profile 1 in Protocol 3.8

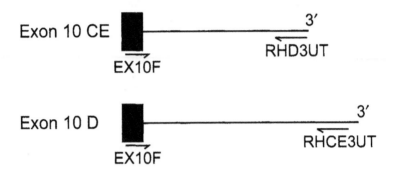

PCR Products

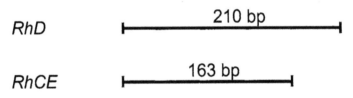

Figure 4.7   *RH* (D/nonD) in exon 10, AS-PCR.

PCR Condition: Cocktail A in Protocol 3.8
Thermal Cycler: Profile 2 in Protocol 3.8

PCR Product |⎯⎯⎯⎯⎯⎯ 218 bp ⎯⎯⎯⎯⎯⎯|

RFLP with *BsaJ* I

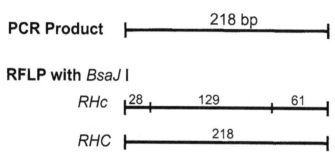

Figure 4.8   *RHC(C)/RHc(c)* in absence of *RHD*, PCR-RFLP.

PCR Condition: Cocktail B in Protocol 3.8
Thermal Cycler: Profile 2 in Protocol 3.8

PCR Product ————————— 474 bp ——————————

RFLP with *Mnl* I

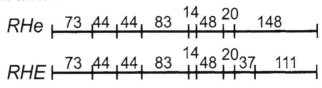

Figure 4.9   *RHE(E)/RHe(e)*, PCR-RFLP.

PCR Condition: Cocktail B or C in Protocol 3.8
Thermal Cycler: Profile 4 in Protocol 3.8

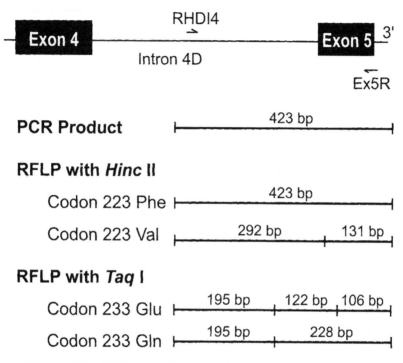

Figure 4.10   *RHD* for certain variants (see Figure 4.11), PCR-RFLP.

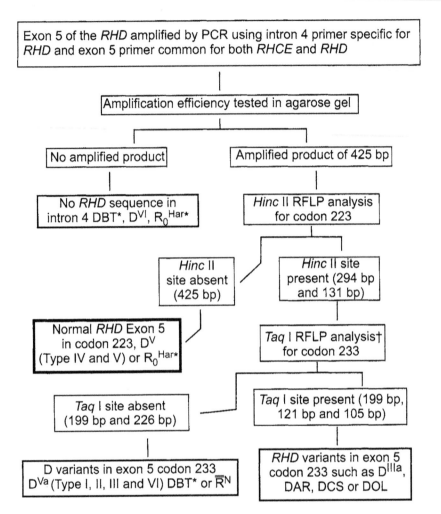

Exon 5 of the *RHD* amplified by PCR using intron 4 primer specific for *RHD* and exon 5 primer common for both *RHCE* and *RHD*

Amplification efficiency tested in agarose gel

No amplified product

Amplified product of 425 bp

No *RHD* sequence in intron 4 DBT*, D^{VI}, R_0^{Har*}

Hinc II RFLP analysis for codon 223

Hinc II site absent (425 bp)

Hinc II site present (294 bp and 131 bp)

Normal *RHD* Exon 5 in codon 223, D^V (Type IV and V) or R_0^{Har*}

Taq I RFLP analysis† for codon 233

Taq I site absent (199 bp and 226 bp)

Taq I site present (199 bp, 121 bp and 105 bp)

D variants in exon 5 codon 233 D^{Va} (Type I, II, III and VI) DBT* or $\overline{R}^N$

RHD variants in exon 5 codon 233 such as D^{IIIa}, DAR, DCS or DOL

* Depending on whether the gene rearrangement includes intron 4 from RHD.
† Using 425 bp PCR product.

Figure 4.11   Analysis and interpretation of PCR-RFLP for *RHD* variants.

Primers

Name	Sequence	GenBank Accession
D/nonD in intron 4:		
RHI41	5'-GTGTCTGAAGCCCTTCCATC-3'	Y10604, Y10605
RHI42	5'-GAAATCTGCATACCCCAGGC-3'	Y10604, Y10605
RHI43	5'-ATTAGCTGGGCATGGTGGTG-3'	Y10605

With RHI41/RHI42 primers for D 115 bp; for CE 715 bp
With RHI41/RHI42/RHI43 primers for D 115 bp; for CE 230 bp*

DIII[a]:		
RHDI4	5'-TAAGCACTTCACAGAGCAGG-3'	Y10604
EX5R	5'-TCTTGCTGATCTTCCCTTGG-3'	X54534, L08429
Rh E/e:		
CEI4	5'-GGCAACAGAGCAAGAGTCCA-3'	Y10605
CEX5	5'-CTGATCTTCCTTTGGGGGTG-3'	X54534, L08429
Rh C/c:		
EX2S	5'-GGCCAAGATCTGACCGTGAT-3'	X54534
EXI2R	5'-TGACCCAGAAGTGATCCAGC-3'	U66340
D/nonD in exon 10:		
EX10F	5'-TTTCCTCATTTGGCTGTTGGATTTTAA-3'	X54534, L08429
RHD3UT	5'-GTATTCTACAGTGCATAATAAATGGTG-3'	(Huang 1996)
RHCE3UT	5'-CTGTCTCTGACCTTGTTTCATTATAC-3'	(Huang 1996)

* The addition of RHI43 to the assay reduces the size of the RHCE target to 230 bp and, which is preferentially amplified over the 715 bp product.

Comments

The antigens are conformation-dependent and although the molecular analyses are focused on a single critical mutation, other polymorphisms are involved in expression of the antigens. Furthermore, numerous rearranged genes encode hybrid proteins.

References

Avent, N.D. and Reid, M.E. (2000) The Rh blood group system: A review. *Blood*, 95, 375–387.

Cartron, J.-P. (1996) Rh DNA – Coordinator's Report. *Transfus Clin Biol* 3, 491–495.

Huang, C.-H. (1996) Alteration of *RH* gene structure and expression in human dCCee and DC[w]- red blood cells: Phenotypic homozygosity versus genotypic heterozygosity. *Blood* 88, 2326–2333.

Huang, C.-H., Liu, P.Z. and Cheng, J.G. (2000) Molecular biology and genetics of the Rh blood group system. *Semin Hematol*, in press.

Wagner, F.F., Gassner, C., Müller, T.H., et al (1999) Molecular basis of weak D phenotypes. *Blood* 93, 385–393.

P Blood Group System
Facts Sheet

ISBT Gene Name:	*P1*
Organization:	Not known
Chromosome:	22q11.2–qter
Gene product:	Galactosyltransferase
GenBank Accession:	Gene has not been cloned.
Alleles:	*P1*
Antigens:	P1

Prevalence of gene products (% based on phenotyping):

Phenotype	Caucasians	Blacks	Cambodians
P1+	79	94	20
P1–	21	6	80

References

Spitalnik, P.F. and Spitalnik, S.L. (1995) The P blood group system: Biochemical, sero
logical, and clinical aspects. *Transf Med Rev* 9, 110–122.

Lutheran Blood Group System
Facts Sheet

ISBT Gene Name: *LU*

Organization: 15 exons distributed over ~12 kbp (cDNA 2,402 bp)

Chromosome: 19q13.2–q13.3

Gene product: Lutheran glycoprotein (597 amino acids)
B-CAM (557 amino acids)

GenBank Accession: X83425

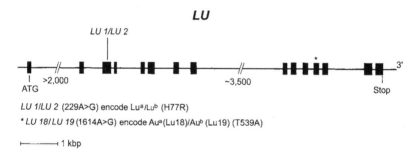

LU 1/LU 2 (229A>G) encode Luᵃ/Luᵇ (H77R)

* *LU 18/LU 19* (1614A>G) encode Auᵃ(Lu18)/Auᵇ (Lu19) (T539A)

├────────┤ 1 kbp

Figure 4.12   *LU* gene.

Alleles: *LU 1/LU 2*

Antigens: Luᵃ/Luᵇ and numerous variants

Prevalence of gene products (% based on phenotyping):

Phenotype	Most populations
Lu(a+b−)	0.2
Lu(a+b+)	7.4
Lu(a−b+)	92.4

Molecular Protocols

PCR Condition: Cocktail B or C in Protocol 3.8
Thermal Cycler: Profile 1 in Protocol 3.8

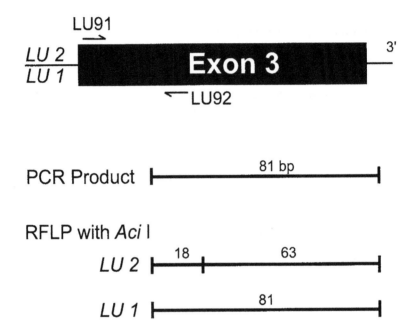

Figure 4.13 *LU 1* (Lu^a)/*LU 2* (Lu^b), PCR-RFLP.

Primers

Name	Sequence	GenBank Accession
LU91	5′-CGCTCGGGAGCTCGCCCC-3′	X83425
LU92	5′-GGCCCCGGGTGTCGTGCA-3′	X83425

References

El Nemer, W., Rahuel, C., Colin, Y., et al (1997) Organization of the human LU gene and molecular basis of the Lu^a/Lu^b blood group polymorphism. *Blood* 89, 4608–4616.

Parsons, S.F., Mallinson, G., Daniels, G.L., et al (1997) Use of domain-deletion mutants to locate Lutheran blood group antigens to each of the five immunoglobulin super-family domains of the Lutheran glycoprotein: Elucidation of the molecular basis of the Lu^a/Lu^b and the Au^a/Au^b polymorphisms. *Blood* 89, 4219–4225.

Parsons, S.F., Mallinson, G., Holmes, C.H., et al (1995) The Lutheran blood group glycoprotein, another member of the immunoglobulin superfamily, is widely expressed in human tissues and is developmentally regulated in human liver. *Proc Natl Acad Sci USA* 92, 5496–5500.

Rahuel, C., Kim, C.L., Mattei, M.G., et al (1996) A unique gene encodes spliceoforms of the B-cell adhesion molecule cell surface glycoprotein of epithelial cancer and of the Lutheran blood group glycoprotein. *Blood* 88, 1865–1872.

Kell Blood Group System
Facts Sheet

ISBT Gene Name:	*KEL*
Organization:	19 exons distributed over 21.5 kbp (cDNA 2,199 bp)
Chromosome:	7q33
Gene product:	Kell glycoprotein (732 amino acids)
GenBank Accession:	M64934, AH004703

KEL

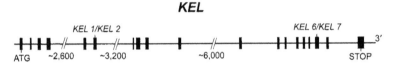

KEL 1/KEL 2 (698T>C) encode K/k (M193T)
KEL 6/KEL 7 (1910C>T) encode Js^a/Js^b (P597L)
For other variants, see Table 4.6

Figure 4.14 *KEL* gene.

Alleles:	*KEL 1/KEL 2; KEL 3/KEL 4; KEL 6/KEL 7*; and numerous variants
Antigens:	K/k, Kp^a/Kp^b, Js^a/Js^b and numerous variants

Prevalence of gene products (% based on phenotyping):

Phenotype	Caucasians	Blacks
K–k+	91	98
K+k–	0.2	rare
K+k+	8.8	2
Kp(a+b–)	rare	0
Kp(a–b+)	97.7	100
Kp(a+b+)	2.3	rare
Js(a+b–)	0	1
Js(a–b+)	100	80
Js(a+b+)	rare	19

Molecular Protocols

PCR Condition: Cocktail C in Protocol 3.8
Thermal Cycler: Profile 1 in Protocol 3.8

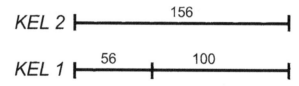

PCR Product |—————— 156 bp ——————|

RFLP with *Bsm* I

KEL 2 |——————— 156 ———————|

KEL 1 |—— 56 ——|——— 100 ———|

Figure 4.15   *KEL 1* (K)/*KEL 2* (k), PCR-RFLP.

PCR Condition: Cocktail A or C in Protocol 3.8
Thermal Cycler: Profile 1 in Protocol 3.8

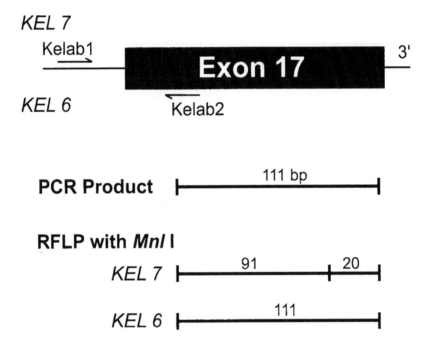

Figure 4.16   *KEL 6* (Js[a])/*KEL 7* (Js[b]), PCR-RFLP.

Primers

Name	Sequence	GenBank Accession
KELS	5'-AAGCTTGGAGGCTGGCGCAT-3'	M64934
KELR	5'-CCTCACCTGGATGACTGGTG-3'	M64934
Kelab1	5'-CTCACCTAGGCAGCACCAACCCTA-3'	(Lee, et al 1995)
Kelab2	5'-TTCCTGGAGGGCATGGTTGTCACA-3'	M64934

Table 4.10

Other variants

Antigen	Amino acid substitution	Exon	Nucleotide mutation	Restriction enzyme
Kp^a/Kp^b	W281R	8	961T>C	*Nla* III (+/–)
Kp^b/Kp^c	R281Q	8	962G>A	*Pvu* II (–/+)
Ul(a–)/Ul(a+)	E494V	13	1601A>T	*Acc* I (–/+)
K11/K17	V302A	8	1025T>C	*Hae* III (–/+)
K14/K24	R180P	6	659G>C	*Hae* III (–/+)
K23–/K23+	Q382R	10	1265A>G	*Bcn* I (–/+)
K12+/K12–	H548R	15	1763A>G	*Nla* III (+/–)
K18+/K18–	R130W	4	508C>T	*Taq* II (–/+)
	R130Q	4	509G>A	*Eco57* I (–/+)
K19+/K19–	R492Q	13	1595G>A	
K22+/K22–	A322V	9	1085C>T	*Tsp45* I (–/+)
Tou+/Tou–	R406Q	11	1337G>A	

(+/–) corresponds to presence or absence of restriction enzyme site

References

Lee, S. (1997) Molecular basis of Kell blood group phenotypes. *Vox Sang* 73, 1–11.

Lee, S., Zambas, E.D., Marsh, W.L., et al (1991) Molecular cloning and primary structure of Kell blood group protein. *Proc Natl Acad Sci USA* 88, 6353–6357.

Lee, S., Wu, X., Reid M., Redman, C. (1995) Molecular basis of k:6,→[Js(a+b–)] phenotype in the kell blood group system. *Transfusion* 35, 822–825.

Lewis Blood Group System
Facts Sheet

ISBT Gene Name: *LE*

Organization: 3 exons distributed over >8 kbp (cDNA 1,086 bp)

Chromosome: 19p13.3

Gene product: α-1-3/1-4-L-fucosyltransferase (361 amino acids)

GenBank Accession: U27328; NM00149; D89324

LE (FUT3)

Figure 4.17 *LE* gene.

Antigens: Le^a and Le^b, depending on the interaction of other gene products

Prevalence of gene products (% based on phenotyping):

Phenotype	Caucasians	Blacks
Le(a+b–)	22	23
Le(a–b+)	72	55
Le(a–b–)	6	22

Comment

Lewis antigens are adsorbed onto red blood cells from plasma. Molecular assays are not given.

Reference

Cameron, H.S., Szczepaniak, D. and Weston, B.W. (1995) Expression of human chromosome 19p α(1,3)-fucosyltransferase genes in normal tissues. Alternative splicing, polyadenylation, and isoforms. *J Biol Chem* 270, 20112–20122.

Duffy Blood Group System
Facts Sheet

ISBT Gene Name: *FY*

Organization: 2 exons distributed over 1.521 kbp (cDNA 1,041 bp)

Chromosome: 1q22–q23

Gene product: Major (β) Duffy glycoprotein (336 amino acids)
Minor (α) Duffy glycoprotein (338 amino acids)

GenBank Accession: U01839; S76830

Alleles: *FY 1/FY 2*

Antigens: Fy^a/Fy^b, Fy3

Prevalence of gene products (based on phenotyping):

Phenotype	Caucasians	Blacks	Chinese	Japanese	Thai
Fy(a+b–)	17	9	90.8	81.5	69
Fy(a–b+)	34	22	0.3	0.9	3
Fy(a+b+)	49	1	8.9	17.6	28
Fy(a–b–)	Rare	68	0	0	0

Comments

The Fy^x phenotype is characterized by low expression of Fy^b. The molecular basis for Fy^x is due to 265C > T and 298G > A in *FYB*.

Fy(a–b–) is associated with 133C > T that disrupts the binding motif for the transcription initiation in the erythroid lineage. In four people the Fy(a–b–) phenotype is associated with nonsense mutation ($n = 3$) or with a deletion of 14 nucleotides ($n = 1$).

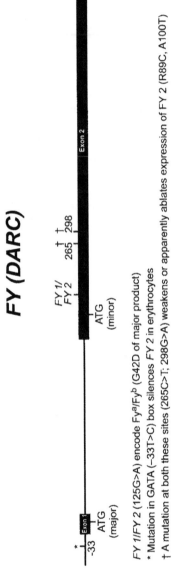

FY 1/FY 2 (125G>A) encode Fyᵃ/Fyᵇ (G42D of major product)

* Mutation in GATA (−33T>C) box silences *FY 2* in erythrocytes

† A mutation at both these sites (265C>T; 298G>A) weakens or apparently ablates expression of *FY 2* (R89C, A100T)

Figure 4.18   *FY* gene.

Molecular Protocols

PCR Condition: Cocktail A or C in Protocol 3.8
Thermal Cycler: Profile 1 in Protocol 3.8

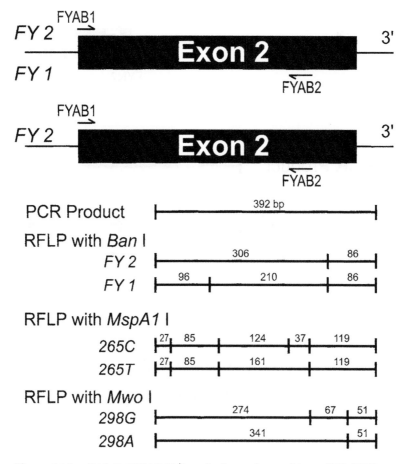

Figure 4.19   *FY 1* (Fy^a)/*FY 2* (Fy^b) and other polymorphisms, PCR-RFLP.

PCR Condition: Cocktail A or C in Protocol 3.8
Thermal Cycler: Profile 1 in Protocol 3.8

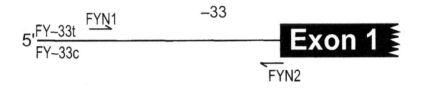

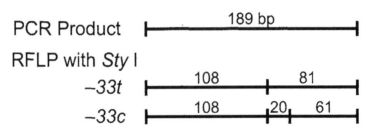

Figure 4.20 FY GATA-1, PCR-RFLP.

Primers

Name	Sequence	GenBank Accession
FYAB1	5'-TCCCCCTCAACTGAGAACTC-3'	S576830
FYAB2	5'-AAGGCTGAGCCATACCAGAC-3'	S576830
FYN1	5'-CAAGGCCAGTGACCCCCATA-3'	S576830
FYN2	5'-CATGGCACCGTTTGGTTCAG-3'	S576830

References

Chaudhuri, A., Polyakova, J., Zbrzezna, V., et al (1995) The coding sequence of Duffy blood group gene in humans and simians: Restriction fragment length polymorphism, antibody and malarial parasite specificities, and expression in non-erythroid tissues in Duffy-negative individuals. *Blood* 85, 615–621.

Iwamoto, S., Omi, T., Kajii, E., et al (1995) Genomic organization of the glycophorin D gene: Duffy blood group Fy^a/Fy^b alloantigen system is associated with a polymorphism at the 44-amino acid residue. *Blood* 85, 622–626.

Mallinson, G., Soo, K.S., Schall, T.J., et al (1995) Mutations in the erythrocyte chemokine receptor (Duffy) gene: The molecular basis of the Fy^a/Fy^b antigens and identification of a deletion in the Duffy gene of an apparently healthy individual with the Fy(a-b-) phenotype. *Br J Haematol* 90, 823–829.

Olsson, M.L., Smythe, J.S., Hansson, C., et al (1998) The Fy^x phenotype is associated with a missense mutation in the Fy^b allele predicting Arg89Cys in the Duffy glyco-protein. *Br J Haematol* 103, 1184–1191.

Parasol, N., Reid, M., Rios, M., et al (1998) A novel mutation in the coding sequence of the FY*B allele of the Duffy chemokine receptor gene is associated with an altered erythrocyte phentoype. *Blood* 92, 2237–2243.

Tournamille, C., Colin, Y., Cartron, J.P., et al (1995) Disruption of a GATA motif in the Duffy gene promoter abolishes erythroid gene expression in Duffy-negative individuals. *Nature Genet* 10, 224–228.

Tournamille, C., Le Van Kim, C., Gane, P., et al (1995) Molecular basis and PCR-DNA typing of the Fy^a/Fy^b blood group polymorphism. *Hum Genet* 95, 407–410.

Tournamille, C., Le, V.K., Gane, P., et al (1998) Arg89Cys substitution results in very low membrane expression of the Duffy antigen/receptor for chemokines in Fy^x individuals. *Blood* 92, 2147–2156.

Kidd Blood Group System
Facts Sheet

ISBT Gene Name: *JK*

Organization: 11 exons distributed over 30 kbp (cDNA 1,596 bp)

Chromosome: 18q11–q12

Gene product: Urea transporter (391 amino acids)

GenBank Accession: L36121

JK (SLC14A1)

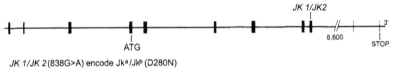

JK 1/JK 2(838G>A) encode Jkª/Jkᵇ (D280N)

⊢—⊣ 1 kbp

Figure 4.21   *JK* gene.

Alleles: *JK 1/JK 2*

Antigens: Jkª/Jkᵇ, Jk3

Prevalence of gene products (based on phenotyping):

Phenotype	Caucasians	Blacks	Asians
Jk(a+b–)	26.3	51.1	23.2
Jk(a–b+)	23.4	8.1	26.8
Jk(a+b+)	50.3	40.8	49.1

Molecular Protocols

PCR Condition: Cocktail A or C in Protocol 3.8
Thermal Cycler: Profile 1 in Protocol 3.8

PCR Product |—————————— 210 bp ——————————|

RFLP with *Mnl* I

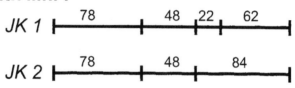

Figure 4.22   *JK 1* (Jkᵃ)/*JK 2* (Jkᵇ), PCR-RFLP.

Primers

Name	Sequence	GenBank Accession
JKIS	5′-TGAGATCTTGGCTTCCTAGG-3′	Unpublished
JK2	5′-ATTGCAATGCAGGCCAGAGA-3′	L36121

References

Lucien, N., Sidoux-Walter, F., Olivès, B., et al (1998) Characterization of the gene encoding the human Kidd blood group/urea transporter protein: Evidence for splice site mutations in Jk$_{null}$ individuals. *J Biol Chem* 273, 12973–12980.

Olivès, B., Neau, P., Bailly, P., et al (1994) Cloning and functional expression of a urea transporter from human bone marrow cells. *J Biol Chem* 269, 31649–31652.

Olivès, B., Mattei, M.-G., Huet, M., et al (1995) Kidd blood group and urea transport function of human erythrocytes are carried by the same protein. *J Biol Chem* 270, 15607–15610.

Olivès, B., Martial, S., Mattei, M.G., et al (1996) Molecular characterization of a new urea transporter in the human kidney. *FEBS Lett* 386, 156–160.

Olivès, B., Merriman, M., Bailly, P., et al (1997) The molecular basis of the Kidd blood group polymorphism and its lack of association with type 1 diabetes susceptibility. *Hum Mol Genet* 6. 1017–1020.

Diego Blood Group System
Facts Sheet

ISBT Gene Name: *DI*

Organization: 20 exons distributed over 228 kbp (cDNA 5,451 bp)

Chromosome: 17q21-q22

Gene product: Anion exchanger 1 (Band 3) (911 amino acids)

GenBank Accession: X77738, M27819

DI (SLC4A1)

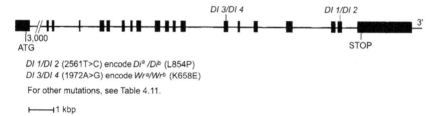

DI 1/DI 2 (2561T>C) encode *Diᵃ /Diᵇ* (L854P)
DI 3/DI 4 (1972A>G) encode *Wrᵃ/Wrᵇ* (K658E)

For other mutations, see Table 4.11.

├───────┤1 kbp

Figure 4.23   *DI* gene.

Alleles: *DI 1/DI 2*; *DI 3/DI 4*; numerous variants

Antigens: Diᵃ/Diᵇ; Wrᵃ/Wrᵇ and numerous low incidence antigens

Prevalence of gene products (% based on phenotyping):

Phenotype	Caucasians	Blacks	Asians	South American Indians
Di(a+b–)	<0.01	<0.01	<0.01	<0.1
Di(a–b+)	>99.9	>99.9	90	64
Di(a+b+)	<0.1	<0.1	10	36
Wr(a+b–)	Rare			
Wr(a–b+)	>99.9			
Wr(a+b+)	<0.01			

Table 4.11

Other variants

Antigen	Amino acid substitution	Exon	Nucleotide mutation	Restriction enzyme
Di^b/Di^a (Memphis II)	P854L	18	2561C>T	Nae I (–/+)
	K56E	3	166A>G	Dra III (–/+)
Wr^b/Wr^a	E658K	15	1972G>A	
Wd(a–)/Wd(a+)	V557M	13	1669G>A	Msl I (–/+)
Rb(a–)/Rb(a+)	P548L	13	1643C>T	EcoN I (–/+)
WARR–/WARR+	T552I	13	1655C>T	Bbs I (+/–)
ELO–/ELO+	R432W	11	1294C>T	Msp I (+/–)
				BstN I (+/–)
Wu–/Wu+	G565A	13	1694G>C	Apa I (+/–)
Bp(a–)/Bp(a+)	N569K	13	1707C>A	Tth2
Mo(a–)/Mo(a+)	R656H	15	1967G>A	Bsm I (–/+)
Hg(a–)/Hg(a+)	R656C	15	1966C>T	Cac8 I (–/+)
Vg(a–)/Vg(a+)	Y555H	13	1666T>C	Dra III (–/+)
Sw(a–)/Sw(a+)	R646Q	15	1937G>A	
BOW–/BOW+	P561S	13	1681C>T	Ban I (+/–)
				BstE II (–/+)
NFLD–/NFLD+	E429D	11	1287A>T	
	P561A	13	1681C>G	
Jn(a–)/Jn(a+)	P566S	13	1696C>T	
KREP–/KREP+	P566A	13	1696C>G	

References

Bruce, L.J., Anstee, D.J., Spring, F.A., et al (1994) Band 3 Memphis variant II. Altered stilbene disulfonate binding and the Diego (Di^a) blood group antigen are associated with the human erythrocyte band 3 mutation Pro[854]→Leu. *J Biol Chem* 269, 16155–16158

Schofield, A.E., Martin, P.G., Spillett, D., et al (1994) The structure of the human red blood cell anion exchanger (EPB3, AE1, band 3) gene. *Blood* 84, 2000–2012.

Zelinski, T. (1998) Erythrocyte band 3 antigens and the Diego blood group system. *Transf Med Rev* 12, 36–45.

Yt Blood Group System
Facts Sheet

ISBT Gene Name: *YT*

Organization: 6 exons; sizes of introns have not been determined (cDNA 2,218 bp)

Chromosome: 7q22.1

Gene product: Acetylcholinesterase (557 amino acids)

Genbank Accession: L42812 (exons 2–6) *also*: L06484 (promoter region, exons 1–2); L22559 (exon 2); L22560 (exon 3); L22561 (exon 4, 5); L22562 (exon 6); M55040 (human acetylcholinesterase mRNA); M76539 (exons 2 and 3H)

YT (ACHE)

ATG   STOP

* *YT 1/YT 2* (1057C>A) encode Yt^a/Yt^b (H353N)

⊢—1 kbp

Figure 4.24   *YT* gene.

Alleles: *YT 1/YT 2*

Antigens: Yt^a/Yt^b

Prevalence of gene products (% based on phenotyping):

Phenotype	Most populations	Israelis
Yt(a+b–)	91.2	97
Yt(a–b+)	Rare	0
Yt(a+b+)	8	23

References

Bartels, C.F., Zelinski, T. and Lockridge, O. (1993) Mutation at codon 322 in the human acetylcholinesterase (ACHE) gene accounts for YT blood group polymorphism. *Am J Hum Genet* 52, 928–936.

Li, Y., Camp, S., Rachinsky, T.L., et al (1991) Gene structure of mammalian acetylcholinesterase: Alternative exons dictate tissue-specific expression. *J Biol Chem* 266, 23083–23090.

Rao, N., Whitsett, C.F., Oxendine, S.M., et al (1993) Human erythrocyte acetylcholinesterase bears the Yta blood group antigen and is reduced or absent in the Yt(a-b-) phenotype. *Blood* 81, 815–819.

Spring, F.A., Gardner, B. and Anstee, D.J. (1992) Evidence that the antigens of the Yt blood group system are located on human erythrocyte acetylcholinesterase. *Blood* 80, 2136–2141.

Xg Blood Group System
Facts Sheet

ISBT Gene Name:	*XG*
Organization:	10 exons distributed over 80 kbp (bone marrow) (cDNA 1,029 bp) 11 exons (fibroblasts)
Chromosome:	Xp22.3
Gene product:	Xg^a glycoprotein (180 amino acids)
GenBank Accession:	S73261

Xg (PBDX)

ATG   STOP

⊢⊣ 1 kbp

Figure 4.25   *XG* gene.

Alleles:	*XG 1*
Antigens:	Xg^a

Prevalence of gene products (% based on phenotyping):

Phenotype	Male	Female
Xg(a+)	66	89
Xg(a−)	34	11

Comments

Part of *XG* is pseudoautosomally linked and part is X-linked. The molecular basis of Xg^a is not yet determined.

References

Ellis, N.A., Ye, T.Z., Patton, S., et al (1994a) Cloning of *PBDX*, an *MIC2*-related gene that spans the pseudoautosomal boundary on chromosome Xp. *Nature Genet* 6, 394–400.

Ellis, N.A., Tippett, P., Petty, A., et al (1994b) PBDX is the XG blood group gene. *Nature Genet* 8, 285–290.

Tippett, P. and Ellis, N.A. (1998) The Xg blood group system: A review. *Transf Med Rev* 12, 233–257.

Scianna Blood Group System
Facts Sheet

ISBT Gene Name: *SC*

Organization: Not known

Chromosome: 1p35–p32

Gene product: Sc glycoprotein

GenBank Accession: Gene has not been cloned

Alleles: *SC 1/SC 2*

Antigens: Sc1/Sc2; Sc3

Prevalence of gene products (% based on phenotyping):

Phenotype	Most populations
Sc:1,–2	99
Sc:–1,2	rare
Sc:1,2	1

Reference

Spring, F.A., Herron, R. and Rowe, G. (1990) An erythrocyte glycoprotein of apparent Mr 60,000 expresses the Sc1 and Sc2 antigens. *Vox Sang* 58, 122–125.

Dombrock Blood Group System
Facts Sheet

ISBT Gene Name:	*DO*
Organization:	Not known
Chromosome:	12p13.2–p12.1
Gene product:	Do glycoprotein
GenBank Accession:	Gene has not been cloned
Alleles:	*DO 1/DO 2; DO 3; DO 4; DO 5*
Antigens:	Do^a/Do^b, Gy^a, Hy, Jo^a

Prevalence of gene products (% based on phenotyping):

Phenotype	Caucasians	Blacks
Do(a+b–)	18	11
Do(a–b+)	33	45
Do(a+b+)	49	44

References

Banks, J.A., Hemming, N. and Poole, J. (1995) Evidence that the Gy^a, Hy and Jo^a antigens belong to the Dombrock blood group system. *Vox Sang* 68, 177–182.

Spring, F.A. and Reid, M.E. (1991) Evidence that the human blood group antigens Gy^a and Hy are carried on a novel glycosylphosphatidylinositol-linked erythrocyte membrane glycoprotein. *Vox Sang* 60, 53–59.

Spring, F.A., Reid, M.E. and Nicholson, G. (1994) Evidence for expression of the Jo^a blood group antigen on the Gy^a/Hy-active glycoprotein. *Vox Sang* 66, 72–77.

Colton Blood Group System Facts Sheet

ISBT Gene Name: CO

Organization: 4 exons distributed over 2.2 kbp (cDNA 816 bp)

Chromosome: 7p14

Gene product: Channel-forming integral protein (CHIP-1) (269 amino acids)

GenBank Accession: D31846, M77829; L11320-L11327

CO (AQP1)

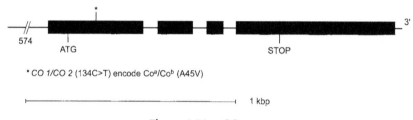

* CO 1/CO 2 (134C>T) encode Co^a/Co^b (A45V)

Figure 4.26   *CO* gene.

Alleles: *CO 1/CO 2*

Antigen: Co^a/Co^b, Co3

Prevalence of gene products (% based on phenotyping):

Phenotype	Most populations
Co(a+b–)	90
Co(a–b+)	0.5
Co(a+b+)	9.5

References

Agre, P., Smith, B.L., Baumgarten, R., et al (1994) Human red cell Aquaporin CHIP. II. Expression during normal fetal development and in a novel form of congenital dyserythropoietic anemia. *J Clin Invest* 94, 1050–1058.

Preston, G.M. and Agre, P. (1991) Isolation of the cDNA for erythrocyte integral membrane protein of 28 kilodaltons: Member of an ancient channel family. *Proc Natl Acad Sci USA* 88, 11110–11114.

Smith, B.L., Preston, G.M., Spring, F.A., et al (1994) Human red cell Aquaporin CHIP. I. Molecular characterization of ABH and Colton blood group antigens. *J Clin Invest* 94, 1043–1049.

Uchida, S., Sasaki, S., Fushimi, K., et al (1994) Isolation of human aquaporin-CD gene. *J Biol Chem* 269, 23451–23455.

Landsteiner–Wiener Blood Group System Facts Sheet

ISBT Gene Name:	*LW*
Organization:	3 exons distributed over 2.6 kbp (cDNA 1,387 bp)
Chromosome:	19p13.3
Gene product:	LW glycoprotein (241 amino acids)
GenBank Accession:	L27670; L27671; X93093; S78852; S78853

LW

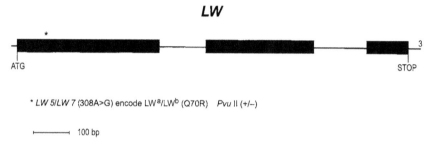

* *LW 5/LW 7* (308A>G) encode LW^a/LW^b (Q70R)   *Pvu* II (+/–)

⊢——————⊣ 100 bp

Figure 4.27   *LW* gene.

Alleles:	*LW 5/LW 7*
Antigens:	LW^a/LW^b; LW^{ab}

Prevalence of gene products (% based on phenotyping):

Phenotype	Most populations	Finns
LW(a+b–)	97	93.9
LW(a–b+)	rare	0.1
LW(a+b+)	3	6.0

References

Bailly, P., Hermand, P., Callebaut, I., et al (1994) The LW blood group glycoprotein is homologous to intercellular adhesion molecules. *Proc Natl Acad Sci USA* 91, 5306–5310.

Bailly, P., Tontti, E., Hermand, P., et al (1995) The red cell LW blood group protein is an intercellular adhesion molecule which binds to CD11/CD18 leukocyte integrins. *Eur J Immunol* 25, 3316–3320.

Casasnovas, J.M., Springer, T.A., Liu, J.-H., et al (1997) Crystal structure of ICAM-2 reveals a distinctive integrin recognition surface. *Nature* 387, 312–315.

Hermand, P., Gane, P., Mattei, M.G., et al (1995) Molecular basis and expression of the LW^a/LW^b blood group polymorphism. *Blood* 86, 1590–1594.

Mizuno, T., Yoshihara, Y., Inazawa, J., et al (1997) cDNA cloning and chromosomal localization of the human telencephalin and its distinctive interaction with lymphocyte function-associated antigen-1. *J Biol Chem* 272, 1156–1163.

Chido/Rodgers Blood Group System
Facts Sheet

ISBT Gene Name:	*CH (C4B)*; *RG (C4A)*
Organization:	C4A 41 exons distributed over 21 kbp (cDNA 5,406 bp)
	C4B 41 exons
Chromosome:	6p21.3
Gene product:	C4A and C4B complement components
GenBank Accession:	K02403; M59815; M59816; U24578
Alleles:	*CH/RG 1* and 6 variants
	CH/RG 11 and 1 variant
Antigens:	Ch1, Ch2, Rg1 and 6 others

Prevalence of gene products (% based on phenotyping):

Phenotype	Most populations
Ch+Rg+	94
Ch+Rg–	2
Ch–Rg+	4
Ch–Rg–	rare

Comment

Chido and Rodgers antigens are expressed on C4, which is adsorbed onto red blood cells from plasma. Ch and Rg antigens are located on C4d, which is a tryptic fragment of C4. Molecular assays are not given.

References

Belt, K.T., Carroll, M.C. and Porter, R.R. (1984) The structural basis of the multiple forms of human complement component C4. *Cell* 36, 907–914.

Giles, C.M. (1988) Antigenic determinants of human C4, Rodgers and Chido. *Exp Clin Immunogenet* 5, 99–114.

Ulgiati, D., Townend, D.C., Christiansen, F.T., et al (1996) Complete sequence of the complement C4 gene from the HLA-A1, B8, C4AQ0, C4B1, DR3 haplotype. *Immunogenetics* 43, 250–252.

Yu, C.Y. (1991) The complete exon-intron structure of a human complement component C4A gene. DNA sequences, polymorphism, and linkage to the 21-hydroxylase gene. *J Immunol* 146, 1057–1066.

Yu, C.Y., Campbell, R.D. and Porter, R.R. (1988) A structural model for the location of the Rodgers and the Chido antigenic determinants and their correlation with the human complement component C4A/C4B isotypes. *Immunogenetics* 27, 399–405.

Hh Blood Group System
Facts Sheet

ISBT Gene Name: *H*

Organization: 4 exons distributed over 8 kbp (cDNA 1,095 bp)

Chromosome: 19q13.3

Gene product: α-2-L-fucosyltransferase (365 amino acids)

GenBank Accession: Z69587

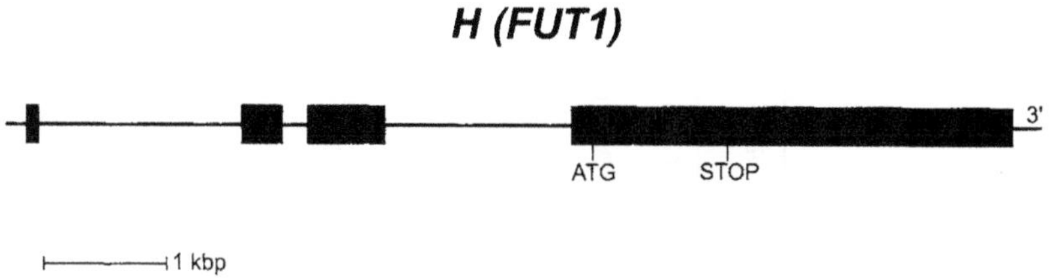

H (FUT1)

Figure 4.28   *H* gene. Most variants are enclosed by missense mutations, see Table 4.12.

Alleles: *H 1*

Antigens: H

Prevalence of gene products (% based on phenotyping):

Phenotype	Most populations
H+	100 (strength variable depending on ABO group)

Comments

Variant forms of *H* give rise to Para-Bombay phenotypes. Deleted *H* gives rise to Bombay (Oh) phenotype

Table 4.12
H(FUT1) variants

Phenotype	Allele	nt: aa:	1	35 12	460 154	547–552	658 220	880–882	980 327	1042 348	1096
O	H1		ATG	C Ala	T Tyr	AGAGAG	C Arg	TTT	A Asn	G Glu	TGA
Para-Bombay	h1		–	–	–	AGAG ↑	–	–	–	–	–
Para-Bombay	h2		–	–	–	–	–	T	–	–	–
Para-Bombay	h3		–	–	–	–	T Cys	↑	–	–	–
Para-Bombay	h4		–	T Val	–	–	–	–	C Thr	–	–
Para-Bombay	h5		–	–	C His	–	–	–	–	–	–
B$_m^h$	–		–	–	C His	–	–	–	–	A Lys	–

References

Koda, Y., Soejima, M. and Kimura, H. (1997) Structure and expression of H-type GDP-L-fucose:β-D-galactoside 2-α-L-fucosyltransferase gene (*FUT1*). Two transcription start sites and alternative splicing generate several forms of *FUT1* mRNA. *J Biol Chem* 272, 7501–7505.

Larsen, R.D., Ernst, L.K., Nair, R.P., et al (1990) Molecular cloning, sequence, and expression of a human GDP-L-fucose:beta-D-galactoside 2-alpha-L-fucosyltransferase cDNA that can form the H blood group antigen. *Proc Natl Acad Sci USA* 87, 6674–6678.

Wang, B., Koda, Y., Soejima, M., et al (1997) Two missense mutations of H type α(1,2)fucosyltransferase. *Vox Sang* 72, 31–35.

Yu, L.-C., Yang, Y.-H., Broadberry, R.E., et al (1997) Heterogeneity of the human H blood group α (1,2) fucosyltransferase gene among para-Bombay individuals. *Vox Sang* 72, 36–40.

Kx Blood Group System
Facts Sheet

ISBT Gene Name: *XK*

Organization: 3 exons. Sizes of introns have not been determined (cDNA 5,096 bp)

Chromosome: Xp21.1

Gene product: Xk glycoprotein (444 amino acids)

GenBank Accession: Z32684

XK

Figure 4.29   *XK* gene.

Alleles: *XK 1*

Antigens: Kx

Prevalence of gene products (% based on phenotyping):

Phenotype	Most populations
Kx+	100

Comments

Deletion of, and variant forms of, *XK* give rise to Kx-negative males who have the McLeod syndrome. No polymorphisms have been described on the expressed XK proteins.

Reference

Ho, M., Chelly, J., Carter, N., et al (1994) Isolation of the gene for McLeod syndrome that encodes a novel membrane transport protein. *Cell* 77, 869–880.

Gerbich Blood Group System
Facts Sheet

ISBT Gene Name:	*GE*
Organization:	4 exons distributed over 13.5 kbp (cDNA 915 bp)
Chromosome:	2q14–q21
Gene product:	Glycophorin C (GPC) (128 amino acids)
	Glycophorin D (GPD) (107 amino acids)
GenBank Accession:	M36284

GE (GYPC)

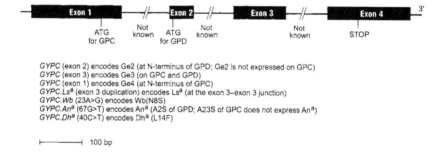

GYPC (exon 2) encodes Ge2 (at N-terminus of GPD; Ge2 is not expressed on GPC)
GYPC (exon 3) encodes Ge3 (on GPC and GPD)
GYPC (exon 1) encodes Ge4 (at N-terminus of GPC)
GYPC.Ls^a (exon 3 duplication) encodes Ls^a (at the exon 3–exon 3 junction)
GYPC.Wb (23A>G) encodes Wb(N8S)
GYPC.An^a (67G>T) encodes An^a (A2S of GPD; A23S of GPC does not express An^a)
GYPC.Dh^a (40C>T) encodes Dh^a (L14F)

├────────┤ 100 bp

Figure 4.30   *GE* genes.

Alleles:	*GE* 2 and 6 variants
Antigens:	Ge2, Ge3, Ge4, Wb, Ls^a, An^a, Dh^a

Prevalence of gene products (% based on phenotyping):

Phenotype	Most populations
Ge2, Ge3, Ge4	100
Wb, Ls^a, An^a, Dh^a	rare

References

Colin, Y., Le Van Kim, C., Tsapis, A., et al (1989) Human erythrocyte glycophorin C. Gene structure and rearrangement in genetic variants. *J Biol Chem* 264, 3773–3780.

Colin, Y., Rahuel, C., London, J., et al (1986) Isolation of cDNA clones and complete amino acid sequence of human erythrocyte glycophorin C. *J Biol Chem* 261, 229–233.

Reid, M.E. and Spring, F.A. (1994) Molecular basis of glycophorin C variants and their associated blood group antigens. *Transf Med* 4, 139–149.

Cromer Blood Group System
Facts Sheet

ISBT Gene Name: *CROM*

Organization: 11 exons distributed over 40 kbp (cDNA 2,102 bp)

Chromosome: 1q32

Gene product: Decay accelerating factor (DAF; CD55) (347 amino acids)

GenBank Accession: M31516, M30142, 35156

CROM (DAF, CD55)

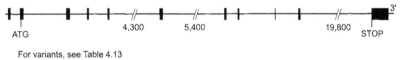

ATG   4,300   5,400   19,800   STOP   3'

For variants, see Table 4.13

├──┤1 kbp

Figure 4.31   *CROM* gene.

Alleles: *CROM 1* and 9 variants

Antigens: Cr^a, Tc^a, Tc^b, Tc^c, Dr^a, Es^a, IFC, WES^a, WES^b, UMC

Prevalence of gene products (% based on phenotyping):

Phenotype	Most populations
Cr^a, Tc^a, Dr^a, Es^a, IFC, WES^b, UMC	100
Tc^b, Tc^c, WES^a	rare

Table 4.13

Cromer Variants

Antigen	Amino acid substitution	Exon	Nucleotide mutation	Restriction enzyme
Cr(a+)/Cr(a−)	A193P	6	679G>C	
Tcᵃ/Tcᵇ	R18L	2	155G>T	*Rsa* I (+/−)
				Stu I (−/+)
Tcᵃ/Tcᶜ	R18P	2	155G>C	*Rsa* I (+/−)
				Stu I (−/−)
Dr(a+)/Dr(a−)	S165L	5	596C>T	*Taq* I (+/−)
Es(a+)/Es(a−)	I46N	2	239T>A	*Sau3* A I (+/−)
IFC+/IFC−	W53 Stop	2	261G>A	*Bcl* I (−/+)
	54 Stop*		263C>A	*Mbo* II (+/−)
WESᵇ/WESᵃ	L48R	2	245T>G	*Afl* II (+/−)
UMC+/UMC−	T216M	6	749C>T	

* Nucleotide mutation produces a cryptic splice site with resultant 26 base pair deletion, frameshift and downstream stop codon.

References

Caras, I.W., Davitz, M.A., Rhee, L., et al (1987) Cloning of decay-accelerating factor suggests novel use of splicing to generate two proteins. *Nature* 325, 545–549.

Daniels, G. (1989) Cromer-related antigens: Blood group determinants on decay-accelerating factor. *Vox Sang* 56, 205–211.

Lublin, D.M. and Atkinson, J.P. (1989) Decay-accelerating factor: Biochemistry, molecular biology and function. *Ann Rev Immunol* 7, 35–58.

Lublin, D.M., Kompelli, S., Storry, J.R., et al (2000) Molecular basis of Cromer blood group antigens. *Transfusion*, in press.

Medof, M.E., Lublin, D.M., Holers, V.M., et al (1987) Cloning and characterization of cDNAs encoding the complete sequence of decay-accelerating factor of human complement. *Proc Natl Acad Sci USA* 84, 2007–2011.

Post, T.W., Arce, M.A., Liszewski, M.K., et al (1990) Structure of the gene for human complement protein decay accelerating factor. *J Immunol* 144, 740–744.

Knops Blood Group System
Facts Sheet

ISBT Gene Name: *KN*

Organization: 39 exons distributed over 140 kbp in A allele (variant form exists with 8 extra exons)

Chromosome: 1q32

Gene product: Complement receptor type 1 (CR1; CD35) (1,998 amino acids)

GenBank Accession: Y00816, X05309; X14893

KN (CR1)

ATG

STOP

3'

⊢ 1 kbp

Figure 4.32   *KN* gene.

Alleles: *KN 1/KN 2* and 3 variants

Antigens: Kn^a/Kn^b, McC^a, Sl^a, Yk^a

Prevalence of gene products (% based on phenotyping):

Phenotype	Caucasians	Blacks
Kn(a+b−)	94.5	99.9
Kn(a−b+)	1	0
Kn(a+b+)	4.5	0.1
McC(a+)	98	94
Sla(a+)	98	60
Yka(a+)	92	98

Comment

The molecular basis of these antigens has not been determined.

References

Ahearn, J.M. and Fearon, D.T. (1989) Structure and function of the complement receptors, CR1 (CD35) and CR2 (CD21). *Adv Immunol* 46, 183–219.

Klickstein, L.B., Wong, W.W., Smith, J.A., et al (1987) Human C3b/C4b receptor (CR1). Demonstration of long homologous repeating domains that are composed of the short consensus repeats characteristics of C3/C4 binding proteins. *J Exp Med* 165, 1095–1112.

Vik, D.P. and Wong, W.W. (1993) Structure of the gene for the F allele of complement receptor type 1 and sequence of the coding region unique to the S allele. *J Immunol* 151, 6214–6224.

Indian Blood Group System
Facts Sheet

ISBT Gene Name: *IN*

Organization: At least 19 exons distributed over 50 kbp (10 exons are variable) (cDNA 1,794 bp)

Chromosome: 11p13

Gene product: CD44 (341 amino acids)

GenBank Accession: M59040; M33827; M69215

IN (CD44)

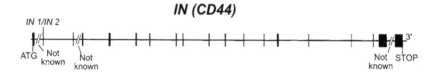

IN 1/IN 2 (252C>G) encode In^a/In^b (P26R)

⊢⊣1 kbp

Figure 4.33   *IN* gene.

Alleles: *IN 1/IN 2*

Antigens: In^a/In^b

Prevalence of gene products (% based on phenotyping):

Phenotype	Caucasians/Blacks	Indians	Iranians/Arabs
In(a+b−)	Rare	Rare	Rare
In(a−b+)	99	96	93
In(a+b+)	<0.1	4	7

References

Goldstein, L.A., Zhou, D.F.H., Picker, L.J., et al (1989) A human lymphocyte homing receptor, the Hermes antigen, is related to cartilage proteoglycan core and link proteins. *Cell* 56, 1063–1072.

Screaton, G.R., Bell, M.V., Jackson, D.G., et al (1992) Genomic structure of DNA encoding the lymphocyte homing receptor CD44 reveals at least 12 alternatively spliced exons. *Proc Natl Acad Sci USA* 89, 12,160–12,164.

Telen, M.J., Udani, M., Washington, M.K., et al (1996) A blood group-related polymorphism of CD44 abolishes a hyaluronan-binding consensus sequence without preventing hyaluronan binding. *J Biol Chem* 271, 7147–7153.

OK Blood Group System
Facts Sheet

ISBT Gene Name: *OK*

Organization: 7 exons distributed over 1.8 kbp (cDNA 810 bp)

Chromosome: 19p13.3

Gene product: CD147 (248 amino acids)

GenBank Accession: AC005559; L10240; X64364

OK (CD147)

6,800

ATG

3'

STOP

* *OK* (331G>A) encodes Ok(a+)/Ok(a−) (E92K)

├────────┤ 1 kbp

Figure 4.34   *OK* gene.

Alleles: *OK 1*

Antigens: Ok^a

Prevalence of gene products (% based on phenotyping):

Phenotype	Most populations
Ok(a+)	100%

References

Biswas, C., Zhang, Y., DeCastro, R., et al (1995) The human tumor cell-derived collagenase stimulatory factor (renamed EMMPRIN) is a member of the immuno-globulin superfamily. *Cancer Res* 55, 434–439.

Guo, H., Majmudar, G., Jensen, T.C., et al (1998) Characterization of the gene for human EMMPRIN, a tumor cell surface inducer of matrix metalloproteinases. *Gene* 220, 99–108.

Spring, F.A., Holmes, C.H., Simpson, K.L., et al (1997) The Oka blood group antigen is a marker for the M6 leukocyte activation antigen, the human homolog of OX-47 antigen, basigin and neurothelin, an immunoglobulin superfamily molecule that is widely expressed in human cells and tissues. *Eur J Immunol* 27, 891–897.

Raph Blood Group System
Facts Sheet

ISBT Gene Name: *MER2*

Organization: Not known

Chromosome: 11p15.5

Gene product: Not defined

GenBank Accession: The gene has not been cloned.

Alleles: *MER2*

Antigens: MER2

Prevalence of gene products (% based on phenotyping):

Phenotype	Most populations
MER2+	92
MER2–	8

Comment

The molecular basis of this antigen is not known.

Reference

Daniels, G.L., Levene, C., Berrebi, A., et al (1988) Human alloantibodies detecting a red cell antigen apparently identical to MER2. *Vox Sang* 55, 161–164.

Part 5   Platelet Blood Groups

5.1   Terminology

Historically, the nomenclature for platelet antigens was based on the name of the proband for which the first antibody was characterized. Therefore, no simple rule linked a given antigen with a gene product. A nomenclature, introduced by the Working Party on Platelets of the International Society of Blood Transfusion (ISBT), simplified the naming of human platelet antigens (HPA) by assigning each antigen a number (von dem Borne and Decary 1990). However, the ISBT system does not indicate which gene product expresses a given platelet antigen. Many of these antigens are antithetical variants of the same platelet plasma membrane constituent, glycoprotein IIIa (GP3A), which is also known as $\beta 3$-integrin. Newman, in an attempt to standardize platelet terminology, discussed the merits of three classification systems to distinguish antigens that are variants of the same gene product (Newman 1994). One system uses amino acid terminology (e.g., Pro_{33} GPIIIa), making it the most informative. However, it has the disadvantage of having to wait for the genetic polymorphism to be characterized. A second modified HPA nomenclature introduced the use of a lower case alphabetical assignment for allelic variants as they are characterized for a given gene product (e.g., HPA-1a, -1b, -1c). The third system proposed by Newman, a gene-based nomenclature (e.g., GPIIIa-01, -02, -03), is similar to the ISBT system for RBC and is consistent with the Guidelines for Genetic Nomenclature of genes and their mutations (Shows et al 1987; White et al 1997; Antonarakis et al 1998; Santoso and Kiefel 1998). For example, the K/k and Kp^a/Kp^b blood group antigens are all products of *KEL*, and by ISBT nomenclature are designated as *KEL 1/KEL 2* and *KEL 3/KEL 4*, respectively. For platelets, the GP3A antigens HPA-1a/HPA-1b, HPA-4b, HPA-6b, HPA-7b and HPA-8b would be named *GP3A 1/GP3A 2,*

GP3A 3, *GP3A 4*, *GP3A 5* and *GP3A 6*. The wild-type GP3A 1 allele encodes a protein that in addition to HPA-1a also carries the high incidence antigens HPA-4a, HPA-6a, HPA-7a and HPA-8a.

In this book, platelet antigen systems are organized by the gene that carries the allelic polymorphisms and that are important in transfusion medicine. The low incidence antigens are tabulated for reference purposes but protocols for their detection are not given.

Table 5.1

Platelet systems

Platelet system		Gene name		Alleles for molecular
Classical	ISBT	ISBT	HGM	protocols given
Br/Zav	HPA-5	*GP1A*	*ITGA2*	*GP1A 1/GP1A 2*
Sit	HPA-13	*GP1A*	*ITGA2*	(*GP1A 1/GP1A 3*)
Ko	HPA-2	*GP1BA*	*GP1BA*	*GP1BA 1/GP1BA 2*
Iy	HPA-12	*GP1BB*	*GP1BB*	(*GP1BB 1/GP1BB 2*)
Bak/Lek	HPA-3	*GP2B*	*ITGA2B*	*GP2B 1/GP2B 2*
Max	HPA-9	*GP2B*	*ITGA2B*	(*GP2B 2/GP2B 3*)
Pl^A/Zw	HPA-1	*GP3A*	*ITGB3*	*GP3A 1/GP3A 2*
Pen/Yuk	HPA-4	*GP3A*	*ITGB3*	*GP3A 1/GP3A 3*
Ca/Tu	HPA-6	*GP3A*	*ITGB3*	(*GP3A 1/GP3A 4*)
Mo	HPA-7	*GP3A*	*ITGB3*	(*GP3A 1/GP3A 5*)
Sr	HPA-8	*GP3A*	*ITGB3*	(*GP3A 1/GP3A 6*)
La	HPA-10	*GP3A*	*ITGB3*	(*GP3A 1/GP3A 7*)
Gro	HPA-11	*GP3A*	*ITGB3*	(*GP3A 1/GP3A 8*)
Oe	HPA-14	*GP3A*	*ITGB3*	(*GP3A 2/GP3A 9*)

Parentheses indicate that the molecular protocol is not given or is not available.

5.2 Clinical Applications

Molecular genotyping is useful to determine the fetal risk for neonatal alloimmune thrombocytopenic purpura (NATP). In particular, molecular genotyping alleviates the need for platelets when the thrombocytopenia is so severe that it is not possible to obtain sufficient platelets for phenotyping.

5.2.1   Identification of a Fetus at Risk for Alloimmune Thrombocytopenia

Analogous to the management of haemolytic disease of the newborn, platelet genotyping is used to determine the paternal antigen inherited by the fetus at risk for NATP. This analysis is relevant when the father is heterozygous for the corresponding allele. Investigators have used chorionic villus, or amniotic fluid-derived DNA to determine the fetal platelet genotype (McFarland et al 1991). These sources avoid the need for fetal blood sampling, which is associated with a higher risk of morbidity or mortality than amniocentesis alone. This risk has been reported to be as high as 10.2% in NATP (Kaplan et al 1994).

We recommend that the decision to perform molecular analysis for clinical use be based on the following:

- The mother has a history of delivery of an infant with NATP.
- The father is heterozygous for the allele or is unknown.
- The genotype is performed on parents and fetus.
- The parental serological phenotypes are determined.
- The ethnicity of the parents is obtained. This can help focus the test approach because some variants are restricted to certain populations (e.g. HPA-4b has a higher incidence in Japanese).
- For the compatible fetus, the genotype is repeated using short-term cultured amniocytes.
- The newborn's platelet antigen genotype is confirmed at birth (cord blood DNA analysis).

Antibodies to antigens in the HPA-1, -3 and -5 systems account for >98% of all confirmed cases of NATP (Warkentin and Smith 1997). However, it is important to recognize that often a population frequency for any given allele is derived from a relatively small number of individuals. The frequency has the potential to represent a bias if obtained from one ethnic group or a geographically isolated area. Therefore, rare alleles may be over-represented or may represent a 'founder allele'.

5.2.2   Genotyping When Appropriate Antisera are not Available

Antibodies to many platelet antigens are not easy to obtain. Furthermore, the platelet surface membrane has a considerable amount of IgG

(2,000–3,000 molecules of IgG/platelet), which complicates the use of serological techniques to determine the phenotype. Thus, serological analysis requires high-titred, well-characterized antisera and the use of 'third generation' techniques (Kiefel et al 1987). In 1989 (Newman et al 1989), the nucleotide polymorphism associated with HPA-1a/-1b was characterized using platelet mRNA and reverse transcription followed by direct PCR sequencing. Since that time, the nucleotide polymorphisms responsible for a number of other antigens expressed on various platelet membrane glycoproteins have been determined. Therefore, molecular techniques have provided the means to type for platelet antigens in the absence of antisera.

5.2.3   Determination of Genotypes in Thrombocytopenic Patients

Molecular genotyping complements serological investigations in the process of antibody identification. This is useful in cases of post-transfusion purpura because it is hard to obtain sufficient platelets for phenotyping. Anti-HPA-1a is the most common antibody associated with the disorder but other antigen systems have been implicated (Waters 1989).

5.2.4   Platelet Panels for Antibody Investigations

Molecular genotyping for platelet antigens can be used to select platelets for panels used in antibody identification. This is possible since inherited defects that cause platelet genotype/phenotype discrepancies are very rare (Skogen and Wang 1996).

5.2.5   Screening for Antigen-negative Platelet Donors

Blood centres should have HPA-1a-negative donors available for treatment of post-transfusion purpura or for NATP. Microtitre assays have been used to identify these donors (Denomme et al 1996; Bessos et al 1996). It is advisable to confirm the phenotype by molecular genotyping.

Furthermore, the DNA can be archived to genotype for other antigens as the need arises; the donor does not need to be recalled for additional testing.

5.3   References

Antonarakis, S.E. and Nomenclature Working Group (1998) Recommendations for a nomenclature system for human gene mutations. *Human Mutation* 11, 1–3.

Bessos, H., Mirza, S., McGill, A., et al (1996) A whole blood assay for platelet HPA1 (PLA1) phenotyping applicable to large-scale screening. *Br J Haematol* 92, 221–225.

Denomme, G., Horsewood, P., Xu, W., et al (1996) A simple and rapid competitive enzyme-linked immunosorbent assay to identify HPA-1a (P1^{A1})-negative donor platelet units. *Transfusion* 36, 805–808.

Kaplan, C., Daffos, F., Forestier, F., et al (1994) Management of fetal and neonatal alloimmune thrombocytopenia. *Vox Sang* 67 (Suppl), 85–88.

Kiefel, V., Santoso, S., Weisheit, M., et al (1987) Monoclonal antibody-specific immobilization of platelet antigens (MAIPA): A new tool for the identification of platelet-reactive antibodies. *Blood* 70, 1722–1726.

McFarland, J.G., Aster, R.H., Bussel, J.B., et al (1991) Prenatal diagnosis of neonatal alloimmune thrombocytopenia using allele-specific oligonucleotide probes. *Blood* 78, 2276–2282.

Newman, P.J. (1994) Nomenclature of human platelet alloantigens: A problem with the HPA system? *Blood* 83, 1447–1451.

Newman, P.J., Derbes, R.S. and Aster, R.H. (1989) The human platelet alloantigen, P1A1 and P1A2, are associated with a leucine33/proline33 amino acid polymorphism in membrane glycoprotein IIIa, and are distinguishable by DNA typing. *J Clin Invest* 83, 1778–1781.

Santoso, S. and Kiefel, V. (1998) Human platelet-specific alloantigens: Update. *Vox Sang* 74 (Suppl 2), 249–253.

Shows, T.B., McAlpine, P.J., Boucheix, C., et al (1987) Guidelines for human gene nomenclature: An international system for human gene nomenclature (ISGN, 1987). *Cytogenet Cell Genet* 46, 11–28.

Skogen, B., Wang, R., McFarland, J.G. and Newman, P.J. (1996) The dinucleotide deletion in exon 4 of the Pl^{A2} allelic form of glycoprotein IIIa: implications for the correlation of serologic versus genotypic analysis of human platelet alloantigens. *Blood* 88, 383–383.

von dem Borne, A.E. and Decary, F. (1990) ICSH/ISBT Working Party on platelet serology: Nomenclature of platelet-specific antigens. *Vox Sang* 58, 176.

Warkentin, T.E. and Smith, J.W. (1997) The alloimmune thrombocytopenic syndromes. *Transf Med Rev* 11, 296–307.

Waters, A.H. (1989) Post-transfusion purpura. *Blood Rev* 3, 83–87.

White, J.A., McAlpine, P.J., Antonarakis, S., et al (1997) Guidelines for human gene nomenclature (1997). *Genomics* 45, 468–471.

5.4   Facts Sheets, Gene Maps and Molecular Protocols

Glycoprotein Ia Platelet Antigens
Facts Sheet

ISBT Gene Name:	*GP1A*
Organization:	Gene not cloned (cDNA 3,546 bp)
Chromosome:	5q23–q31
Gene product:	Glycoprotein Ia (1,181 amino acids)
GenBank Accession:	X17033; M28249 Genomic sequence not available *GP1A 1/GP1A 2* encodes HPA-5a/5b (E505K) *GP1A 1/GP1A 3* encodes HPA-13a/13b (T799M)
Alleles:	*GP1A 1/GP1A 2; GP1A 1/GP1A 3*
Antigens:	HPA-5a/b (Zav$^{b/a}$, Br$^{b/a}$), HPA-13bw (Sit^a)

Prevalence of gene products (% based on genotyping):

Genotype	Cauc.	Black	Chin.	Dutch	Finn	Indo.	Korean
HPA-5a/5a	79	62	68	81	90.2	91	95
HPA-5a/5b	19	34	29	18	9.5	9	5
HPA-5b/5b	2	4	3	1	0.3	–	–
HPA-13b	Low incidence antigen (Caucasian)						

Molecular Protocols

PCR Condition: Cocktail D in Protocol 3.8
Thermal Cycler: Profile 3 in Protocol 3.8

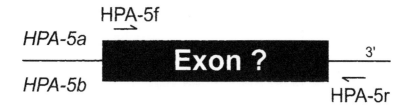

PCR Product   240 bp

RFLP with *Mnl* I

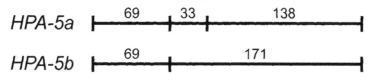

Figure 5.1   *GP1A* (HPA-5), PCR-RFLP.

Primers

Name	Sequence	GenBank Accession
HPA5f	5'-GTTGATGTGGATAAAGACACC-3'	M28249
HPA5r	5'-ATGATGAAATGTAAACCATAC-3'	(intron)

Comments

HPA-5a (*GP1A 1*) has been designated as Zav[b] or Br[b].
  HPA-13a has sequence identity with HPA-5a (*GP1A 1*).

The *GP1A* genomic sequence is unknown. Intronic information for HPA-5 genotyping was obtained by cloning a portion of the cDNA gene (Reiner et al 1998).

References

Reiner, A.P., Aramaki, K.M., Teramura, G., et al (1998) Analysis of platelet glycoprotein Ia (α_2 integrin) allele frequencies in three North American populations reveals genetic association between nucleotide 807C/T and amino acid 505 Glu/Lys (HPA-5) dimorphisms. *Thromb Haemost* 80, 449–456.

Santoso, S., Kalb, R., Walka, M., et al (1993) The human platelet alloantigens Br[a] and Br[b] are associated with a single amino acid polymorphism on glycoprotein Ia (integrin subunit α2). *J Clin Invest* 92, 2427–2432.

Takada, Y. and Hemler, M.E. (1989) The primary structure of the VLA-2/collagen receptor α_2 subunit (platelet GPIa): Homology to other integrins and the presence of a possible collagen-binding domain. *J Cell Biol* 109, 397–407.

Glycoprotein Ibα Platelet Antigens Facts Sheet

ISBT Gene Name:	*GP1BA*
Organization:	2 exons distributed over 6 kbp (cDNA 1,881 bp)
Chromosome:	17p12–pter
Gene product:	Glycoprotein Ibα (626 amino acids)
GenBank Accession:	M22403, D85894 (Exon 2), J02940, S34439 (tandem repeat)

GP1BA

GP1BA 1/GP1BAP 2

ATG

3'

STOP

GP1BA 1/GP1BA 2 C (524C>T) encodes HPA-2a/2b (T145M)

⊢——————————————⊣ 1 kbp

Figure 5.2   *GP1BA* gene.

Alleles:	*GP1BA 1/GP1BA 2*
Antigens:	HPA-2a/b (Ko[b/a], Sib[a])

Prevalence of gene products (% based on genotyping):

Genotype	Cauc.	Black	Amerind.	Chin.	Dutch	Korean
HPA-2a/2a	85	67	91.8	91	87	80
HPA-2a/2b	12	30	8	8.8	12.5	19
HPA-2b/2b	3	3	0.2	2	0.5	1

Molecular Protocols

PCR Condition: Cocktail D in Protocol 3.8
Thermal Cycler: Profile 3 in Protocol 3.8

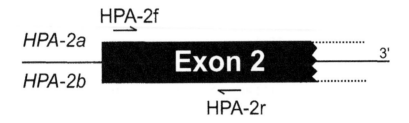

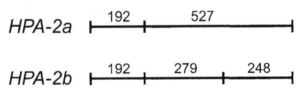

Figure 5.3   *GP1BA* (HPA-2), PCR-RFLP gene.

Primers

Name	Sequence	GenBank Accession
HPA-2f	5'-ATAAGCTTGCTGCTCCTGCTGCCAAGCC-3'	M22403
HPA-2r	5'-TAGAATTCAGCATTGTCCTGCAGCCAGCG-3'	M22403

Comments

HPA-2b (*GP1BA* 2) is designated as Ko[a], Sib[a].

   HPA-2 is associated with a 13 amino acid tandem repeat (Simsek et al 1994).

References

Lopez, J.A., Chung, D.W., Fujikawa, K., et al (1987) Cloning of the alpha chain of human platelet glycoprotein Ib: A transmembrane protein with homology to leucine-rich alpha 2-glycoprotein. *Proc Natl Acad Sci USA* 84, 5615–5619.

Simsek, S., Faber, N.M., Bleeker, P.M., et al (1993) Determination of human platelet antigen frequencies in the Dutch population by immunophenotyping and DNA (allele-specific restriction enzyme) analysis. *Blood* 81, 835–840.

Simsek, S., Bleeker, P.M., van der Schoot, C.E., et al (1994) Association of a variable number of tandem repeats (VNTR) in glycoprotein Iba and HPA-2 alloantigens. *Thromb Haemost* 72, 757–761.

Wenger, R.H., Kieffer, N., Wicki, A.N., et al (1988) Structure of the human blood platelet membrane glycoprotein Ib alpha gene. *Biochem Biophys Res Commun* 156, 389–395.

Glycoprotein Ibβ Platelet Antigens Facts Sheet

ISBT Gene Name: *GP1BB*

Organization: 2 exons distributed over 40 kbp (cDNA 621 bp)

Chromosome: 22q11.2

Gene product: Glycoprotein Ibβ (206 amino acids)

GenBank Accession:

GP1BB

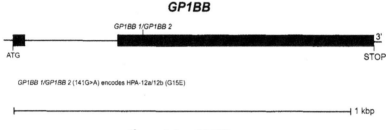

GP1BB 1/GP1BB 2 (141G>A) encodes HPA-12a/12b (G15E)

Figure 5.4   *GP1BB* gene.

Alleles: *GP1BB 1/GP1BB 2*

Antigens: HPA-12bw (Iyᵃ)

Prevalence of gene products (% based on phenotyping):

Phenotype	Caucasians
HPA-12b	Low incidence antigen

Reference

Santoso, S., Böhringer, M., Sachs, U., et al (1996). Point mutation in human platelet glycoprotein Ibβ is associated with the new placelet specific alloantigen Iy(a). Abstract. *Blood* 88 (Suppl 1), 319a.

Yagi, M., Edelhoff, S., Disteche, C.M., et al (1994) Structural characterization and chromosomal location of the gene encoding human platelet glycoprotein Ib beta. *J Biol Chem* 269, 17424–17427.

Glycoprotein IIb Platelet Antigens
Facts Sheet

ISBT Gene Name:	*GP2B*
Organization:	30 exons distributed over 16 kbp (cDNA 3,120 bp)
Chromosome:	17q21.32
Gene product:	Glycoprotein IIb (1,039 amino acids)
GenBank Accession:	M33319; M33320; M34480; J02764
Alleles:	*GP2B 1/GP2B 2; GP2B 2/GP2B 3*
Antigens:	HPA-3a/b (Bak$^{a/b}$, Lek^a), HPA-9bw (Max^a)

Prevalence of gene products (% based on genotyping):

Genotype	Caucasian	Black	Chinese	Dutch	Indonesian	Korean
HPA-3a/3a	46	40	32	31	19	36
HPA-3a/3b	42	45	49	49	53	51
HPA-3b/3b	12	15	19	20	27	13
HPA-9b	Low incidence antigen (Caucasian)					

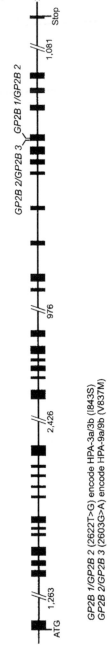

Figure 5.5   *GP2B* gene.

Molecular Protocols

PCR Condition: Cocktail D in Protocol 3.8
Thermal Cycler: Profile 3 in Protocol 3.8

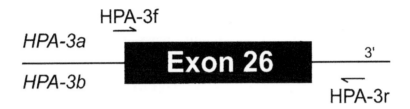

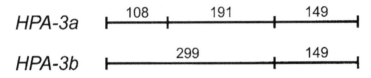

PCR Product 448 bp

RFLP with *Fok* I

HPA-3a   108   191   149

HPA-3b   299   149

Figure 5.6   *GP2B* (HPA-3), PCR-RFLP.

Primers

Name	Sequence	GenBank Accession
HPA3f	5'-TGGAAGAAAGACCTGGGAAGG-3'	M33320
HPA3r	5'-CTCCTTAACGTACTGGGAAGC-3'	M33320

References

Heidenreich, R., Eisman, R., Surrey, S., et al (1990) Organization of the gene for platelet glycoprotein IIb. *Biochemistry* 29, 1232–1244.

Simsek, S., Faber, N.M., Bleeker, P.M., et al (1993) Determination of human platelet antigen frequencies in the Dutch population by immunophenotyping and DNA (allele-specific restriction enzyme) analysis. *Blood* 81, 835–840.

Glycoprotein IIIa Platelet Antigens
Facts Sheet

ISBT Gene Name:	*GP3A*
Organization:	15 exons distributed over 46 kbp (cDNA 2,364 bp)
Chromosome:	Glycoprotein IIIa (788 amino acids)
Gene product:	17q21.32
GenBank Accession:	M32666-86; M57481-94; M20311; J02703
Alleles:	*GP3A 1/GP3A 2; GP3A 1/GP3A 3* and other variant alleles
Antigens:	HPA-1a/b (Pl$^{A1/A2}$/Zw$^{a/b}$), HPA-4a/b (Pen$^{a/b}$/Yuk$^{b/a}$) and additional low incidence variants (see Table 5.2)

Prevalence of gene products (% based on genotyping):

Genotype	Cauc.	Black	Chin.	Dutch	Indo.	Jap.	Kor.
HPA-1a/1a	80	84	99.8	71.6	98.2	99.7	98.5
HPA-1a/1b	18	16	0.2	26	1.8	0.3	1.2
HPA-1b/1b	2	–	–	2.4	–	–	–
HPA-4a/4a	99.8	100	–	100	99.4	97.9	99
HPA-4a/4b	0.2	Rare	–	rare	0.6	2.1	1
HPA-4b/4b	–	–	–	–	–	–	–

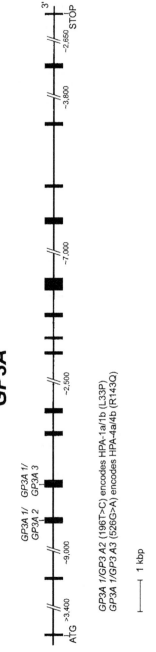

Figure 5.7 *GP3A* gene.

PCR Condition: Cocktail D in Protocol 3.8
Thermal Cycler: Profile 3 in Protocol 3.8

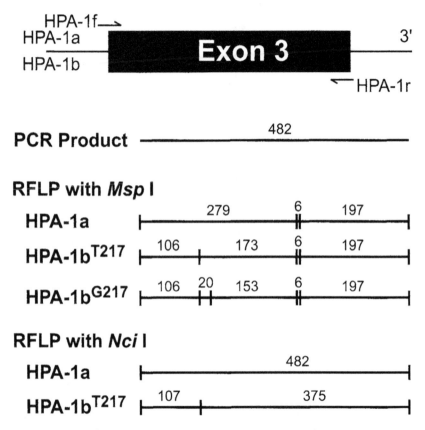

Figure 5.8 *GPIIIA* (HPA-1), PCR-RFLP analysis.

Molecular Protocols

PCR Condition: Cocktail D in Protocol 3.8
Thermal Cycler: Profile 3 in Protocol 3.8

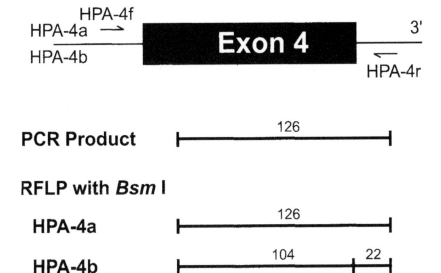

Figure 5.9 *GPIIIA* (HPA-4), PCR-RFLP analysis.

Primers

Name	Sequence	GenBank Accession
HPA1f	5′-CTTAGCTATTGGGAAGTGGTA-3′	M32672
HPA1r	5′-CTTCTGACTCAAGTCCTAACG-3′	M32672
HPA4f	5′-CCTGTGGACATCTACTACTTGATGGACC-3′	M32673
HPA4r	5′-GCCAATCCGCAGGTTACTGGTGAGCATT-3′	M32673

Table 5.2

Other GP3A antigens (variant allele frequency <0.03)

Antigen	Amino acid substitution	Exon	Nucleotide mutation	Restriction enzyme
HPA-6bw (Ca^a/Tu^a)	R489Q	10	1564G>A	Mva I (+)
HPA-7bw (Mo^a)	P407R	10	1317C>G	Bsp1286 I (+)
HPA-8bw (Sr^a)	R636C	12	2004C>T	None
HPA-10bw (La^a)	R62Q	3	283G>A	Ava I (−)
HPA-11bw (Gro^a)	R633H	12	1996G>A	Mae I (−)
HPA-14bw (Oe^a)	L611~	11	1929AAG>del	None

Comments

HPA-4a has sequence identity with HPA-1a (*GP3A 1*).

HPA-4b (GP3A 3) is designated as Pen[b] and Yuk[a].

The original published sequence for genomic GP3A does not include the signal peptide and 5' untranslated region (Fitzgerald et al 1987). Exon assignments were based on the information available at the time. GenBank Accession numbers M32666 through M32686 include the complete gene and list the signal peptide as exon 1. Therefore the exon/intron assignments of the GenBank submission differ from the original publication by +1.

References

Fitzgerald, L.A., Steiner, B., Rall, S.C., Jr., et al (1987) Protein sequence of endothelial glycoprotein IIIa derived from a cDNA clone. Identity with platelet glycoprotein IIIa and similarity to 'integrin'. *J Biol Chem* 262, 3936–3939.

Matsuo, K. and Reid, D.M. (1996) Allele-specific restriction analysis of human platelet antigen system 4. *Transfusion* 36, 809–812.

Newman, P.J., Derbes, R.S. and Aster, R.H. (1989) The human platelet alloantigen, P1[A1] and P1[A2], are associated with a leucine[33]/proline[33] amino acid polymorphism in membrane glycoprotein IIIa, and are distinguishable by DNA typing. *J Clin Invest* 83, 1778–1781.

Simsek, S., Faber, N.M., Bleeker, P.M., et al (1993) Determination of human platelet antigen frequencies in the Dutch population by immunophenotyping and DNA (allele-specific restriction enzyme) analysis. *Blood* 81, 835–840.

Unkelbach, K., Kalb, R., Breitfeld, C., et al (1994) New polymorphism on platelet glycoprotein IIIa gene recognized by endonuclease *Msp* I: implications for Pl[A] typing by allele-specific restriction analysis. *Transfusion* 34, 592–595.

Walchshofer, S., Ghali, D., Fink, M., et al (1994) A rare leucine[40]/arginine[40] polymorphism on platelet glycoprotein IIIa is linked to the human platelet antigen 1b. *Vox Sang* 67, 231–234.

Wang, R., Furihata, K., McFarland, J.G., et al (1992) An amino acid polymorphism within the RGD binding domain of platelet membrane glycoprotein IIIa is responsible for the formation of the Pena/Penb alloantigen system. *J Clin Invest* 90, 2038–2043.

Zimrin, A.B., Gidwitz, S., Lord, S., et al (1990) The genomic organization of platelet glycoprotein IIIa. *J Biol Chem* 265, 8590–8595.

Glycoprotein IV Platelet Antigens Facts Sheet

ISBT Gene Name: *GP4*

Organization: 15 exons distributed over 32 kpb (cDNA 1,419 bp)

Chromosome: 7q11.2

Gene product: Glycoprotein IV (472 amino acids)

Genbank Accession: Z32752-65, Z32770; L06850

Alleles: *GP4 1*

Antigens: Nak^a

Prevalence of gene products (% based on phenotyping):

Phenotype	Caucasians	Blacks	Asians	Dutch
Nak^{a-}	0	2.4	3	0
Nak^{a+}	100	97.6	97	100

References

Armesilla, A.L. and Vega, M.A. (1994) Structural organization of the gene for human CD36 glycoprotein. *J Biol Chem* 269, 18985–18991.

Lipsky, R.H., Sobieski, D.A., Tandon, N.N., et al (1991) Detection of GPIV (CD36) mRNA in Naka- platelets [letter]. *Thromb Haemost* 65, 456–457.

Yamamoto, N., Ikeda, H., Tandon, N.N., et al (1990) A platelet membrane glycoprotein (GP) deficiency in healthy blood donors: Naka- platelets lack detectable GPIV (CD36). *Blood* 76, 1698–1703.

Part 6   Neutrophil Blood Groups

6.1   Terminology

Neutrophil-specific antigens can be elucidated using molecular techniques similar to those used for RBC and platelet antigens. Polymorphisms have been identified on the neutrophil Fc-gamma receptor (FcγR) IIIb glycoprotein, CD11a, CD11b, and on a 58–64 kDa unidentified glycoprotein. The antigen systems described include NA1/NA2, and SH on FcγRIIIb, Ond^a and Mart^a on the β2-integrins, respectively CD11a and CD11b, and NB1 on the 58–64 kDa glycoprotein. The complete absence of FcγRIIIb represents the NA 'null' phenotype and these individuals can develop isoantibodies after pregnancy or transfusion. Genotyping for these polymorphisms requires the generation of cDNA by

Table 6.1

Neutrophil antigens

Neutrophil system		Gene name		Alleles for molecular
Classical	ISBT	ISBT	HGM	protocols given
NA	NA	*NA*	FCGR3B	*NA1/NA2*
NB	NB	*NB*	None	(*NB1/NB2*)
SH	None	None	FCGR3B	*SH⁻/SH⁺*
Ond^a	None	None	*ITGAL*	(*Ond^a/Ond^b*)
Mart^a	None	None	*ITGAM*	(*Mart^a/Mart^b*)
5	5	None	None	(*5a/5b*)
9a	None	None	None	
SL	None	None	None	

Parentheses indicate that the molecular protocol is not given or is not available.
The antithetical 9b and SL⁻ antigens and alleles have not been characterized.

reverse transcription (Simsek et al 1996). The antigens defined by a few remaining antisera have not been characterized nor have the genes been identified (e.g., 5b, 9a and SL). A nomenclature system has not been formally adopted by the ISBT (Bux et al 1997).

6.2   Clinical Applications

Molecular genotyping of neutrophil allelic polymorphisms can be applied to two disorders; neonatal immune neutropenia and transfusion-related acute lung injury (TRALI). The polymorphisms on FcγRIIIb (NA1/NA2 and SH) and on the αL and αM subunits of the β2-integrins (Ond[a] and Mart[a], respectively) have been elucidated at the molecular level.

6.2.1   Immune Neutropenia

Antibodies to the neutrophil antigens NA1 and NA2 have been demonstrated in autoimmune and alloimmune neutropenia. The identification of neutrophil-specific antibodies is important to distinguish congenital neonatal neutropenia or sepsis from immune neutropenia. In neonatal alloimmune neutropenia, a woman develops an antibody to an antigen expressed on neonatal neutrophils inherited from the father but absent on her neutrophils. This disorder is analogous to HDN and NATP. However, it is relatively benign and usually resolves spontaneously.

6.2.2   Transfusion-Related Acute Lung Injury (TRALI)

TRALI is characterized by lung oedema after a blood transfusion and can be fatal (Kopko and Holland 1999). The onset of the disorder is usually acute and rapid. It has been demonstrated that anti-neutrophil agglutinating antibodies are associated with the disease. The antibodies are either produced by the recipient with specificity for donor neutrophil antigens or are found in the plasma of the donor unit and are specific for antigens on the recipient neutrophils. Both anti-NA1 and anti-NB1 have been implicated in TRALI (Stroncek 1997).

6.3   References

Bux, J. and Chapman, J. (1997) Report on the Second International Granulocyte Serology Workshop. *Transfusion* 37, 977–983.

Kopko, P.M. and Holland, P.V. (1999) Transfusion-related acute lung injury. *Br J Haematol* 105, 322–329.

Simsek, S., van der Schoot, C.E., Daams, M., et al (1996) Molecular characterization of antigenic polymorphisms (Ond^a and Mart^a) of the β_2 family recognized by human leukocyte alloantisera. *Blood* 88, 1350–1358.

Stroncek, D.F. (1997) Granulocyte immunology: Is there a need to know? Transfusion 37, 886–888.

6.4   Facts Sheets, Gene Maps and Molecular Protocols

Neutrophil Antigens
Facts Sheet

ISBT Gene Name: *NA (FCGR3B)*

Organization: 5 exons, genomic size unknown (cDNA 702 bp)

Chromosome: 1q23

Gene product: Fcγ receptor IIIb (233 amino acids)

GenBank Accession: Z46223; M90743-6; X16863; J04162

NA (FCGR3B)

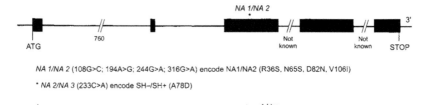

NA 1/NA 2 (108G>C; 194A>G; 244G>A; 316G>A) encode NA1/NA2 (R36S, N65S, D82N, V106I)

* *NA 2/NA 3* (233C>A) encode SH–/SH+ (A78D)

Figure 6.1   *NA* gene.

Alleles: *NA 1/NA 2, NA 3*

Antigens: NA1/NA2, SH+

Prevalence of gene products (% based on phenotyping):

Phenotype	Cauc.	Native Blacks	Americans	Chinese	Jap.	Kor.
NA1/NA1	12	na	na	48	42	na
NA1/NA2	46	na	na	42	45	na
NA2/NA2	42	na	na	10	13	na
SH⁻	4.5	22.5	1.1	na	na	0

Molecular Protocols

PCR Condition: Cocktail in Protocol 3.10
Thermal Cycler: Profile 2 in Protocol 3.10

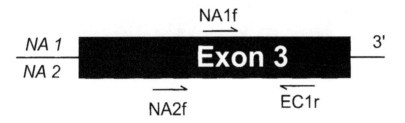

AS-PCR Products

Figure 6.2 *NA 1* (FCGR3B 1)/*NA 2* (FCGR3B 2), AS-PCR.

PCR Condition: Cocktail in Protocol 3.10
Thermal Cycler: Profile 2 in Protocol 3.10

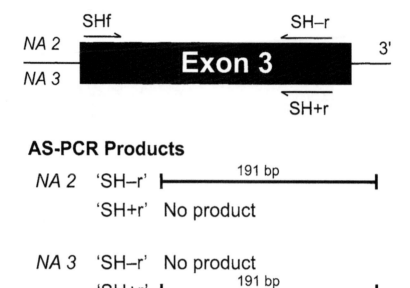

AS-PCR Products

NA 2 'SH–r' |———————— 191 bp ————————|

'SH+r' No product

NA 3 'SH–r' No product

'SH+r' |———————— 191 bp ————————|

Internal control (439 bp with HGH primers) must be present in all reactions.

Figure 6.3 *NA 2 (SH–)/NA 3 (SH+)*, AS-PCR.

Primers

Name	Sequence	Genbank Accession
NA1f	5'-CAGTGGTTTCACAATGTGAA-3'	Z46223
NA2f	5'-CAATGGTACAGCGTGCTT-3'	Z46223
NAr	5'-ATGGACTTCTAGCTGCAC-3'	J04162
SHf	5'-AAGATCTCCCAAAGGCTGTG-3'	Z46223
SH+r	5'-ACTGTCGTTGACTGTGTCAT-3'	J04162
SH-r	5'-ACTGTCGTTGACTGTGTCAG-3'	J04162

Internal control primer

HGH-I	5'-CAGTGCCTTCCCAACCATTCCCTTA-3'	V00520
HGH-II	5'-ATCCACTCACGGATTTCTGTTGTGTTTC-3'	V00520

Comments

SH+ is associated with gene duplication (Koene et al 1998).

References

Bux, J., Stein, E.L., Santoso, S., et al (1995) NA gene frequencies in the German population, determined by polymerase chain reaction with sequence-specific primers. *Transfusion* 35, 54–57.

Koene, H.R., Kleijer, M., Roos, D., et al (1998) *FcγRIIIB* gene duplication: Evidence for presence and expression of three distinct *FcγRIIIB* genes in NA(1+,2+)SH(+) individuals. *Blood* 91, 673–679.

Ory, P.A., Clark, M.R., Kwoh, E.E., et al (1989) Sequences of complementary DNAs that encode the NA1 and NA2 forms of Fc receptor III on human neutrophils. *J Clin Invest* 84, 1688–1691.

Ravetch, J.V. and Perussia, B. (1989) Alternative membrane forms of Fc gamma RIII(CD16) on human natural killer cells and neutrophils. Cell type-specific expression of two genes that differ in single nucleotide substitutions. *J Exp Med* 170, 481–497.

Simsek, S., van der Schoot, C.E., Daams, M., et al (1996) Molecular characterization of antigenic polymorphisms (Ond^a and Mart^a) of the β_2 family recognized by human leukocyte alloantisera. *Blood* 88, 1350–1358.

Part 7   Appendices

7.1   Database Information

Most institutions have access to the World Wide Web (WWW) via the Internet. A plethora of URL (Uniform Resource Locator) sites exist that have useful background information for molecular genetics and molecular biology protocols. The list below is not meant to be exhaustive. Furthermore, some of the sites listed here are mirrored on other web sites. The URL addresses listed below are current at the time of going to press. However, often the web site administrators may update the page or change the address.

www.bioc.aecom.yu.edu/bgmut/index.htm

Blood Group Antigen Mutation Database. A mutation database of gene loci encoding common and rare blood group antigens. Albert Einstein College of Medicine of Yeshiva University in New York.

www.ncbi.nlm.nih.gov/Entrez/

A browser provided by the National Center for Biotechnology Information (NCBI) that queries the GenBank Sequence Database and related MEDLINE resources.

www.ncbi.nlm.nih/BLAST/

NCBI's sequence similarity search tool for nucleotides and proteins.

www.ncbi.nlm.nih/Omim/

Online Mendelian Inheritance in Man (OMIM). A database of human genes and genetic disorders containing textual and reference information. It also contains links to Entrez and MEDLINE.

www.hgmp.mrc.ac.uk/gdb/

The Genome Database contains descriptions of genes, PCR markers, maps of the human genome, and lists of genetics variations, including mutations and polymorphisms. Hosted by The Hospital for Sick Children, Toronto, ON.

www.uct.ac.za/microbiology/manualin.html

Department of Microbiology, University of Cape Town, South Africa. Molecular biology protocols for PCR and polyacrylamide gel electrophoresis.

research.nwfsc.noaa.gov/protocols.html

The Northwest Fisheries Science Center of the US Department of Commerce. General molecular biology protocols and useful information; e.g., error rate of thermal resistant polymerases, an oligonucleotide calculator, and chart of the standard genetic code.

rebase.neb.com

New England BioLabs collection of restriction enzyme database. The web site has a search tool to query the database by enzyme name or recognition sequence.

www.nccls.org/

Home page for the National Committee for Clinical Laboratory (NCCLS) Standards. NCCLS consensus standards, guidelines and best practices available by subscription.

cap.org/

The College of American Pathologists (CAP) is a medical society of pathologists and is considered the leader in providing laboratory quality improvement programmes.

7.2   Glossary

Allele – One of several alternative forms of a gene at a given locus.
Amplification – Production of additional copies of a given DNA.

Annealing – To cause complementary single strands of nucleic acid to pair and form double-stranded DNA or DNA/RNA hybrid.

Base pair (bp) – In DNA two bases, one purine (A, G) and one pyrimidine (T, C), lying opposite each other and joined by hydrogen bonds (complementarity is exclusive A to T and G to C).

Chromosome – A structure comprised of DNA and protein found in the nucleus of a cell. Each chromosome contains hundreds to thousands of genes. Humans have 23 pairs of chromosomes totalling over 50,000 to 100,000 genes.

Codon – A sequence of three nucleotides (triplet) in DNA or RNA that codes for a certain amino acid or for the transcription termination signal.

Complementary DNA (cDNA) – A single strand of DNA synthesized to complement the bases in messenger RNA. Messenger RNA represents the transcriptional product of a gene that is responsible for the production of a protein when translated.

Denaturation – Separation of double-stranded nucleic acid into single strands.

Deoxyribonucleic acid (DNA) – The genetic material of a chromosome. It is a large, double-stranded, helical molecule made of nucleotides.

DNA polymerase – DNA-synthesizing enzyme. To begin synthesis, an oligonucleotide primer is required.

Elongation – Addition of nucleotides to an oligonucleotide strand.

Endonuclease – One of a heterogeneous group of enzymes that cleave bonds between nucleotides of single- or double-stranded DNA or RNA.

Exon – A segment of DNA that is represented in the mature mRNA of eukaryotes.

Exonuclease – An enzyme that cleaves nucleotide chains at their terminal bonds only.

Gene – A hereditary factor that constitutes a single unit of hereditary material. It corresponds to a segment of DNA that codes for the synthesis of a single polypeptide chain.

Gene bank – A collection of cloned DNA fragments that together represent the genome from which they are derived (gene library).

Genome – All of the genetic material of a cell or of an individual.

Genotype – All or a particular part of the genetic constitution of an individual or a cell.

Hybridization – Fusion of two single complementary DNA strands (DNA/DNA hybridization) or of a single strand of DNA and a complementary RNA strand (DNA/RNA hybridization).

Introns – Non-coding DNA sequences that interrupt exons. It is transcribed but removed from the primary RNA transcript before translation.

Locus – The specific place on a chromosome where a gene is located.

Mutation – A permanent inheritable change in DNA sequence. This can be an insertion or deletion or any number of changes or alterations to the gene.

Nucleotide – Single monomeric building block of a polypeptide chain that makes up nucleic acid. A nucleotide is a phosphate ester consisting of a pyrimidine (C or T) or a purine (A or G) base, a sugar (ribose for RNA or deoxyribose for DNA), and a phosphate group.

Oligonucleotide – A short stretch of single-stranded DNA, often used as a probe to find a matching sequence of DNA or RNA or as primers to copy (by polymerase elongation) sequences of DNA or RNA.

Polymerase chain reaction (PCR) – Technique for *in vitro* propagation (amplification) of a given DNA sequence. It is a repetitive thermal cyclic process consisting of denaturation of genomic DNA, annealing the specific oligonucleotide primers to the complementary DNA, and extension of the primers to form the DNA replica.

Phenotype – The observable effect of one or more genes on an individual or a cell.

Polymorphism – A naturally occurring variation in a DNA sequence. Polymorphisms are useful as genetic markers because they distinguish between DNA of different origins. Some polymorphisms in a gene result in a change in the expressed polypeptide.

Restriction site – Enzymes that can recognize specific base sequences and cut DNA into fragments.

Restriction fragment length polymorphism (RFLP) – The production of DNA fragments of different lengths by a given restriction enzyme.

Single-stranded DNA – DNA can be separated into two strands. Separation occurs when genes are being transcribed or when DNA is replicated prior to cell division. Single-stranded DNA can be made in the laboratory with heat during PCR, during the 'melting or denaturation'.

7.3   References

Alberts, B., Bray, D., Lewis, J., et al (1994) *Molecular Biology of the Cell*, 3rd edition. Garland Publishing, New York.

Lewin, B. (1997) *Genes VI*. Oxford University Press, Oxford.

Passarge, E. (1995) *Color Atlas of Genetics*. Georg Thieme Verlag, Stuttgart, New York.

Sambrook, J., Fritsch, E.F. and Maniatis, T. (1989) *Molecular Cloning: A Laboratory Manual* (3 volumes), 2nd edition. Cold Spring Harbor Laboratory Press, Cold Spring Harbor, NY.

7.4 Work Sheets and Request Forms

PCR worksheet

Locus: _____   Thermal Cycler: _____   Date:

SOP: _____   Primers: _____ _____ _____

Lot #: _____ _____ _____

Tube #	Sample ID (Last, First)	DNA (μg/mL)	Vol. (μL) in Reaction	Results (see gel photograph)
1				
2				
3				
4				
5				
6				
7				
8				
9				
10				
11				

Amplification Profile: Use below for calculations

Denaturation _____°C _____ min   10X PCR Buffer Lot#: _____

Cycling steps 1 ____°C _____ sec   MgCl₂ Lot#:_____

2 ____°C _____ sec ____X   dNTPs Lot#:_____

3 ____°C _____ sec   *Taq* Polymerase Lot#: _____

Extension _____°C   10 min   Comments: _____

Soak _____°C hold

Restriction Endonuclease Digestion: Use below for calculations

10X Buffer   2   μL   RE Lot #: _____

5 U RE _____ [name] _____ μL   Buffer Lot#: _____

Sterile ddH₂O _____ μL   Comments:_____

PCR product   8   μL

Total volume _____20   μL   Digested at_____°C for 3 hrs

Reviewed by: _____

Date:

PCR cocktail calculations

Analyte	Volume (µL) one reaction	Volume (µL) × (n + 1) reactions
10X PCR Buffer	5	
MgCl$_2$ (if necessary)		
dNTPs (@ ____ mM)		
Sense Primer	1	
Antisense Primer	1	
Sense Internal Control (AS-PCR only)	(1)	
Antisense Internal Control (AS-PCR only)	(1)	
Taq polymerase		
Sterile ddH$_2$O		
DNA		
TOTAL	50	

Initials: _____

Restriction enzyme cocktail calculations

Analyte	Volume (µL) one reaction	Volume (µL) × (n + 1) reactions
10X Buffer	2	
5 units RE _____ (name)		
Sterile ddH$_2$O		
PCR Product	8	
TOTAL	20	

[Institution]
[address]
[Telephone/Fax/E-mail]

Molecular Genotyping for Blood Groups

Test Requested _____

Type of Sample _____   Date of Collection _____

Demographic Information for Patient

Last Name _____ First Name _____ Middle Initial _____

Hospital I.D. Number _____   Social Security Number _____

Date of Birth _____ Diagnosis _____

Ethnicity (check all that apply):

   Asian☐ Black☐ Hispanic☐ Native American☐ Pacific Islander ☐ White ☐ Other ☐

Clinical Information for Patient

Transfusions in the last 3 months? Number _____ Date(s): _____

24 hours ☐ #_____ 7 days ☐ #_____ 30 days ☐

Stem cell transplantation   Yes ☐   No ☐

Reason for genotyping:

_____
_____

Phenotype:

_____
_____

Comments:

_____
_____

Information on Hospital/Institution Submitting Sample

Hospital/Clinic Name _____ Telephone Number (____)_____

Hospital Address _____ Fax Number (optional) (____)_____

City _____ State _____ Zip Code _____ E-mail (optional) _____

Contact Person _____

Ship overnight delivery at room temperature / 4°C

[Institution]
[address]
[Telephone/Fax/E-mail]

Molecular Genotyping on Amniotic Fluid

Test Requested _____

Demographic Information for Mother

Last Name _____   First Name _____   Middle Initial _____

Hospital I.D. Number _____   Social Security Number

Date of Birth

Ethnicity of Mother (check all that apply):

   Asian☐ Black☐ Hispanic☐ Native American☐ Pacific Islander ☐ White ☐ Other ☐

Ethnicity of Father (check all that apply):

   Asian☐ Black☐ Hispanic☐ Native American☐ Pacific Islander ☐ White ☐ Other ☐

Clinical Information

Has either parent been transfused or transplanted? Yes☐ / No☐

Is this pregnancy a result of non-spousal insemination? Yes☐ / No☐

Samples Required

Amniocytic fluid: (at least 5 ml)   Date of collection _____

Mother's blood: Date of collection _____   Phenotype _____   Antibody _____ Titre _____
                                                           Antibody _____ Titre _____
                                                           Antibody _____ Titre _____

Father's blood: Name _____   Date of collection _____   Phenotype _____

Information on Hospital/Institution Submitting Sample

Hospital/Clinic Name _____   Telephone Number: (____)_____

Hospital Address _____   Fax Number: (____)_____

City _____   State _____   Zip Code _____

Contact Person: Name _____

Ship overnight delivery at room temperature.

Index

Note: Page numbers in **bold** refer to Tables

Useful Tables

Red cell blood group systems

Blood Group System			Gene Name		Molecular
Classical	ISBT No.	ISBT Symbol	ISBT	HGM	protocols provided
ABO	001	ABO	*ABO*	*ABO*	
MNS	002	MNS	*MNS*	*GYPA*	*MNS 1/MNS 2*
				GYPB	*MNS 3/MNS 4*
P	003	P1	*P1*	*P1*	
Rh	004	RH	*RH*	*RHD*	*RH 1*
				RHCE	*RH 3/RH 5*
Lutheran	005	LU	*LU*	*LU*	*LU 1/LU 2*
Kell	006	KEL	*KEL*	*KEL*	*KEL 1/KEL 2*
					KEL 6/KEL 7
Lewis*	007	LE	*LE*	*FUT3*	
Duffy	008	FY	*FY*	*DARC*	*FY 1/FY 2 GATA1;*
					nt265; nt298
Kidd	009	JK	*JK*	*SLC14A1*	*JK 1/JK 2*
Diego	010	DI	*DI*	*SCL4A1*	
Yt	011	YT	*YT*	*ACHE*	
Xg	012	XG	*XG*	*XG*	
Scianna	013	SC	*SC*	*SC*	
Dombrock	014	DO	*DO*	*DO*	
Colton	015	CO	*CO*	*AQP1*	
Landsteiner–Wiener	016	LW	*LW*	*LW*	
Chido/Rodgers*	017	CH/RG	*CH/RG*	*C4A, C4B*	
Hh	018	H	*H*	*FUT1*	
Kx	019	XK	*XK*	*XK*	
Gerbich	020	GE	*GE*	*GYPC*	
Cromer	021	CROM	*CROM*	*DAF*	
Knops	022	KN	*KN*	*CR1*	
Indian	023	IN	*IN*	*CD44*	
Ok	024	OK	*OK*	*CD147*	
Raph	025	RAPH	*MER2*	*MER2*	

* Lewis and Chido/Rogers are not included in section 4.5

Platelet systems

Platelet system		Gene name		Alleles for molecular
Classical	ISBT	ISBT	HGM	protocols given
Br/Zav	HPA-5	*GP1A*	*ITGA2*	*GP1A 1/GP1A 2*
Sit	HPA-13	*GP1A*	*ITGA2*	*(GP1A 1/GP1A 3)*
Ko	HPA-2	*GP1BA*	*GP1BA*	*GP1BA 1/GP1BA 2*
Iy	HPA-12	*GP1BB*	*GP1BB*	*(GP1BB 1/GP1BB 2)*
Bak/Lek	HPA-3	*GP2B*	*ITGA2B*	*GP2B 1/GP2B 2*
Max	HPA-9	*GP2B*	*ITGA2B*	*(GP2B 2/GP2B 3)*
Pl^A/Zw	HPA-1	*GP3A*	*ITGB3*	*GP3A 1/GP3A 2*
Pen/Yuk	HPA-4	*GP3A*	*ITGB3*	*GP3A 1/GP3A 3*
Ca/Tu	HPA-6	*GP3A*	*ITGB3*	*(GP3A 1/GP3A 4)*
Mo	HPA-7	*GP3A*	*ITGB3*	*(GP3A 1/GP3A 5)*
Sr	HPA-8	*GP3A*	*ITGB3*	*(GP3A 1/GP3A 6)*
La	HPA-10	*GP3A*	*ITGB3*	*(GP3A 1/GP3A 7)*
Gro	HPA-11	*GP3A*	*ITGB3*	*(GP3A 1/GP3A 8)*
Oe	HPA-14	*GP3A*	*ITGB3*	*(GP3A 2/GP3A 9)*

Parentheses indicate that the molecular protocol is not given or is not available.

Neutrophil antigens

Neutrophil system		Gene name		Alleles for molecular
Classical	ISBT	ISBT	HGM	protocols given
NA	NA	*NA*	*FCGR3B*	*NA1/NA2*
NB	NB	*NB*	None	*(NB1/NB2)*
SH	None	None	*FCGR3B*	*SH⁻/SH⁺*
Ond^a	None	None	*ITGAL*	*(Ond^a/Ond^b)*
Mart^a	None	None	*ITGAM*	*(Mart^a/Mart^b)*
5	5	None	None	*(5a/5b)*
9a	None	None	None	
SL	None	None	None	

Parentheses indicate that the molecular protocol is not given or is not available.
The antithetical 9b and SL⁻ antigens and alleles have not been characterized.

Printed and bound by CPI Group (UK) Ltd, Croydon, CR0 4YY

03/10/2024

01040413-0013